BUILDING
ADDITIONS

FROM THE EDITORS OF **Fine Homebuilding**®

The Taunton Press

The Taunton Press
Inspiration for hands-on living®

The Taunton Press, Inc., 63 South Main Street, PO Box 5506, Newtown, CT 06470-5506
e-mail: tp@taunton.com

Distributed by Publishers Group West

Cover Design: Cathy Cassidy
Interior Design Layout: Cathy Cassidy
Front Cover Photographer: Andy Engel, courtesy *Fine Homebuilding,* © The Taunton Press, Inc.
Back Cover Photographer: (clockwise from top left) Steve Culpepper, courtesy *Fine Homebuilding,* © The Taunton Press, Inc.;
Andrew Wormer, courtesy *Fine Homebuilding,* © The Taunton Press, Inc.; Tom O'Brien and David Ericson, courtesy *Fine Homebuilding,* © The Taunton Press, Inc.; Roe A. Osborn, courtesy *Fine Homebuilding,* © The Taunton Press, Inc.

For Pros By Pros® is a trademark of The Taunton Press, Inc.,
registered in the U.S. Patent and Trademark Office.

Library of Congress Cataloging-in-Publication Data
Building additions.
       p. cm. -- (For pros, by pros)
   "From the editors of Fine homebuilding."
   ISBN 1-56158-699-4
   1. Buildings--Additions. I. Fine homebuilding. II. Series.
   TH4816.2B85 2004
   690'.8--dc22

                    2004017065

Printed in the United States of America
10 9 8 7 6 5 4 3 2 1

The following manufacturers/names appearing in *Building Additions* are trademarks:
Baldwin® Hardware, Bass® ale, Cultured Stone®, Dec-Klip®, Delta®, Ducksback® Total Wood Finish, Duracoat®, Durajoint® waterstop, Fastap® self drilling screws, Fuller O'Brien® Paints), Fypon Molded Millwork®, Gold Bond® Building Company, Medex®, Minwax® Jacobean, Mr. Potato Head®, National Casein® Co., Olympic WaterGuard®, Panel Lift® (from Telpro Inc.), Pella® Corporation, Sikaflex®, Simpson™, Styrofoam™, Tyvek®, Vise-Grip®, Wal-Mart®, Woodstone® Company

PART 4: CASE STUDIES: INNOVATIVE ADDITIONS

INTRODUCTION

Mosquitos weren't on my mind when I climbed up on the porch roof to replace an old layer of roll roofing with a new one, but the bugs soon discovered me. Without the protection of insect repellent, the feast was on. But I had brought along a secret weapon: two gallons of blind nail cement. This sticky asphalt emulsion is essential for adhering roll roofing to the roof deck and to itself. Soon I was covered with the stuff. And with every well-directed slap, my face, neck, and upper body began to take on a leopard-like appearance. I discovered that the tar spots were immune to insect attack, and I noted that the goo captured and doomed the bugs careless enough to land on it.

Fortunately, the work of winterizing the porch—my first addition project—proceeded more smoothly after the roofing episode. Today, countless construction projects later, I've been able to improve my skills and my design sense by paying attention to *Fine Homebuilding's* expert authors. They're among the best professionals in the building business—veteran contractors and designers with the experience and know-how to guide you through a broad range of addition projects—from building basic foundations and porches to framing dormers and hanging drywall.

For any addition project to succeed, good design is just as important as sound construction. That's why you'll also find an inspiring selection of case studies included here. These projects demonstrate creative, cost-effective ways to build additions and convert existing parts of the house into living space. If you're aiming to make a small house bigger, the ideas and expert advice on these pages will help you achieve topnotch results.

—Tim Snyder
executive editor, *Fine Homebuilding*

An Addition Foundation

■ BY RICK ARNOLD

I recently poured a foundation for a fussy builder. After the forms were stripped, he checked the foundation. The length and width were okay. But the foundation was 1¼ in. out of square, and it was 1 in. out of level from one end to the other. The builder paused for a moment and proclaimed the foundation perfect. The foundation was for an addition to an old house.

A new-house foundation and an addition foundation are two different things. With a new house, I start with an empty hole in the ground and follow the plans: straight, square, plumb, and level. But with an addition foundation, the existing house and its foundation play a big part in molding my strategy.

Sizing Up the Existing Foundation

When I'm planning an addition foundation, the first thing I check is the existing foundation. Here in New England, we encounter three basic types of foundations: poured concrete, stone, and what I call a modified-pier foundation. Poured concrete is the easiest to add to. The walls are usually straight and smooth, and there is a good chance

that the foundation is somewhat square and level.

Stone foundations, standard fare in this area before 1900, are the most challenging to add to. It's a lot of extra work to match concrete forms to the irregular surface of a stone foundation. Stone foundations usually move a great deal over their long lives, so you can bet that they're well out of square and level.

With the project shown in this article, the house was set on concrete piers poured into holes dug in the ground. Over the years, the spaces between the piers had been filled in with block that was not on footings. As is typical, this foundation, as well as the house on it, was neither square nor level.

The Existing Frame Guides the Addition Foundation

Once I've looked at the existing foundation and determined where the addition foundation ties into it, I'm ready to begin layout. I locate the addition foundation according to the frame of the existing house to allow the addition framing to match seamlessly with the house framing (photo, facing page).

Old house meets new foundation. The house dictates the shape and dimensions of the addition foundation. Following the house framing, the string in the foreground locates the footings and the base of the forms, and it guides the final adjustment before the pour.

The short explanation is that I run two sets of strings: one above grade off the mudsills of the house and the other on the floor of the excavation. The type of addition and its attachment to the house determine the degree of difficulty of this process.

If the addition is to be along one wall and away from its ends, such as for a simple bump-out, I set up a string parallel to the existing wall for the addition's outside wall and then square back to the house for the sidewalls.

For the addition in this article, the foundation began at one corner of the house, with the wall of the addition continuing in the same plane as the house wall. If that plane wasn't maintained, then the transition between the old house and the new would have become apparent.

Layout Comes from the Original House

The strings from the old house guide the foundation layout. For a seamless addition, the new framing must align with the old. The new foundation makes or breaks this alignment. A string, parallel to the existing mudsill, stretches along the house to a stake on the other side of the hole (photo, below). From that line, a plumb line first establishes the foundation corner where it intersects with the house on the excavation floor (photo, right) and then the outside corner of the foundation (bottom right photo).

Corner of the House

Line from the House

Outside Corner

I start laying out foundations of this type by extending a string along the mudsill of the existing house and over to the opposite side of the foundation hole (left photo, facing page). There are usually obstructions along the house (in this case, a deck nailer), so I block the string out from the frame (3 in. for this project).

With that string anchored in place, I plumb down to the floor of the excavation at the corner of the house (top right photo, facing page) and at the outside corner of the addition (bottom right photo, facing page), remembering to compensate for the 3-in. offset of the top string. I stretch a string between these two points and then set a third string for the outside wall of the addition, which I set parallel to the house by measuring out from each end.

Square Isn't Necessarily Best

Before going any further, I check to see if these two lines are square to each other. Some builders use a 3-4-5 triangle method. But I prefer the Pythagorean theorem so that I know how far out of square the layout is. I square the measurements of the two sides and add them together. The square root of that number should equal the length of the diagonal. I was pleased to find that the diagonal on this foundation was off less than 1 in. (top photo).

At this point, I go over options with the contractor. Moving the line that runs along the plane of the existing wall usually is not an option. I can square the addition by moving one end of the parallel line in or out, but then the ends of the addition will be different lengths. Or I can leave the lines as is, which keeps all the measurements equal but results in corners that are not 90°.

The contractor's answer usually depends on the details of the addition. In this case, the addition was for a kitchen, and the contractor wanted 90° corners for the cabinets. If the addition had been for bedrooms or for

Sometimes you just don't want to know. Diagonal measurements check the layout and see how far out of square the house is. This house was within an inch of square.

Marking the corners. Rebar pins are used to mark the corners of smaller walls.

a family room, the builder may have preferred to work with equal measurements and out-of-square corners. Sometimes I'm asked to split the difference.

Once the decision is made, I adjust the lines accordingly and finalize the layout on the floor of the excavation. Working from the main addition foundation lines, I locate the corners of all the smaller walls and drive flat foundation rods or ½-in. rebar pins in these spots. I run strings between these pins, outlining the new foundation, then check the smaller walls for square. When satisfied with the layout, I spray-paint the pins orange for visibility.

Foundation layout is set in concrete. When the footings are finished, the layout is transferred from the top string directly to the concrete surface of the footings and checked again for square. Snapped lines then guide the placement of the forms.

Footings Cast the Layout in Concrete

If the top string is going to be in the way while I'm moving and setting the footing forms, I keep the stakes that hold the string in place and temporarily remove the string.

Once the foundation is outlined, I set and pour the footings in the usual fashion. If the footings meet the existing foundation, I tie them together with ½-in. rebar pounded into holes drilled in the existing footing.

After the footings are poured and stripped, I set up the top line again and repeat the process of laying out the foundation (photo, left). This time, instead of driving pins into the soil, I mark the top of the footing with a pencil, then snap chalklines between the marks.

The First Forms Go Against the Existing Foundation

Before I form the walls, I drill a series of ½-in. holes into the existing wall where it will intersect with the new wall. I drill one hole for every 18 in. of wall height. Rebar driven into these holes ties the new foundation to the old (photo, left). I drill the holes about 6 in. deep and let the rebar extend into the new formwork about 18 in. Some jurisdictions may require you to drill over-size holes and to epoxy the rebar in place, but a tight, dry fit satisfies our local code. When I'm tying an addition foundation to a stone foundation, the unevenness of the stone is usually enough to lock the two together.

For the foundation on a new house, I begin setting forms at one corner. With an addition foundation, though, I always begin where it matches up with the existing foundation. For the first set of forms, I measure in the thickness of the foundation (in this case, 10 in.) from the snapped line and mark the footing for the inside form. Whenever

Connecting the old with the new. Rebar driven into drilled holes every 18 in. ties the addition foundation to the old foundation.

Setting and Bracing the Forms

The first forms go in next to the corner of the house (photo, below). A 2x4 scrap nailed to the siding provides a place to tack the inside form plumb. Special foundation rods then connect the free ends of the forms, holding them apart at the proper width. Filling voids between the forms and the original foundation can be as simple as inserting pieces of Styrofoam (top right photo), but often, it requires a collection of wood and other materials. To keep short walls in place during the pour, they are cross-braced from the inside (middle photo), and they are braced against the foundation hole from the outside (bottom right photo).

Filling Voids

Bracing Walls Inside

First Forms

Bracing Walls Outside

possible, I make a plumb mark on the house foundation up from the mark on the footing. I then line up the first form with the two marks and drive a masonry nail through the form and into the old wall to tack the form in place.

Because of the irregular block foundation on this project, I didn't have a place to tack the form, so I nailed a scrap of 2x4 to the house above the sills and tacked the inside form to the 2x4 (left photo, p. 9). I use the same method when adding to stone. Special jump or hook rods that join the free ends of the forms have to be installed on the first form before it goes into place. The masonry nails or the nails through the 2x4 just keep the form from moving laterally under its own weight; the rods keep the forms together as the concrete is being poured.

Once the inside form is set and plumbed, I set the outside form, which is automatically spaced 10 in. from the inside form by the hook rods (top right photo, p. 9). Steel wedges that are installed in the hook rods lock the outside panel in place, so there is no need to nail it to anything.

Rarely does the formwork match up with the existing foundation without there being voids to fill. Large deep voids can be filled with a scribed piece of plywood. However, I usually put the forms against the foundation and then fill the large voids using anything from pieces of wood to Styrofoam™ (top right photo, p. 9).

Once the first forms are set against the old foundation, I measure and set the remaining forms. I never know the exact length of the formwork until the first forms are set, so I bring different-width forms and spacers to make up any odd distance.

I set and brace all the forms as I would with any new foundation. Then I restretch the reference lines between the original stakes over the tops of the forms. The forms now can be adjusted to their proper position, taking into account the 3-in. offset from the existing wall (photo p. 5).

Bracing Walls Inside and Out

At this point, the builder establishes where the top of the pour, or grade, will be. For this project, he cut away a section of existing wall where the new-addition floor would match up with the existing floor. From that point, we subtracted all the elements of the floor and framing to determine the grade of the foundation. Because concrete usually settles a bit right after the pour, I add a good ⅛ in., and then I shoot the grade inside the forms with a transit. I snap lines on the inside of forms and drive grade nails every few feet along the lines.

If the addition floor ties into the existing floor at more than one place or along a large area, I find the lowest point of the existing floor to establish the grade. If the existing house is extremely out of level, I'm sometimes asked to make the new foundation match the error.

At the other end of the addition shown here, the new foundation returned to the old with a short wall. Without bracing, the pressure of the wet concrete could have pushed the wall away from the existing foundation. I had to brace the forms to keep them from kicking out (middle photo, p. 9).

For this brace, I first placed a tall 2x12 footing plank vertically against the bank of the excavation opposite the end of the wall and another against the form. Next, I cut pieces of 2x stock that fit tightly between the two 2x12s and nailed them into position every 2 ft. or 3 ft. I also cross-braced the small walls from inside the foundation, running the bracing back to the opposite footing (bottom photo, p. 9).

Shoring Up Excavated Soil

A common problem with a full cellar-addition foundation is having to dig below the level of the existing foundation. The best way to deal with this problem is to excavate as closely as possible to the existing foundation without undermining it and then to pour a single-face concrete wall (bottom photo), which keeps the soil under the foundation intact and undisturbed.

After the addition walls are poured and stripped, I set up forms about 8 in. away from the part of the excavation that sticks out the farthest. The new walls provide something substantial to brace the single-face wall against.

Because concrete exerts a lot of force as it is being poured, I brace the wall every 2 ft. vertically and every 2 ft. to 3 ft. horizontally (top photo). I shoot the grade to extend 1 in. to 2 in. above the bottom of the existing foundation. The concrete mix should be stiff and should be poured slowly; it also should be poked with the shovel to prevent defects.

You can't have too much bracing. To resist the pressure of wet concrete, the forms are braced every few feet top and bottom, with the addition walls providing a strong backer for the braces.

A single-face wall keeps the soil where it belongs. Excavation for the full basement addition left soil exposed below the existing foundation. The concrete wall poured against the soil keeps the foundation from being undermined.

Filling the Forms

Unless directed otherwise, I use a 2,500-lb. concrete mix, but I order a fairly stiff mix of concrete at a slump of 3 or 4. Slump is a scale of 1 to 12, with 1 being the stiffest, driest mix and 12 being the wettest. The stiff concrete mix plugs any small voids where the new foundation ties into the existing foundation (photo, p. 12).

I began the pour on this project at the corner where I began the layout. As the concrete fills the first forms, I gently poke at it with a 2x4 to help the concrete form to and bond with the existing foundation. A slow rate of pour and an extra pair of eyes help to prevent any mishaps at this juncture.

If access for the concrete trucks is good, I try to alternate between ends and fill the forms in smaller lifts. Depending on the strength of the existing foundation, I initially fill the forms about one-half full to three-quarters full before moving to the other wall that ties into the foundation and repeating the process. As a safety precaution, I then switch back and forth between the two ends, filling a little more each time. Alternating the pour this way gives the concrete a little time to begin to set up, which helps to reduce the initial pressure.

A stiff mix starts the pour. The pour starts at the existing foundation, and a stiff mix helps to plug any small voids where the forms meet the old foundation.

With this project, access to the foundation wasn't great, and I needed a couple of loads of concrete to complete the pour. This fact changed my approach, and I just worked from one end to the other. After the first end was about two-thirds full, I added a small amount of water to the concrete, bringing it to about a 5 slump, to make it flow into the rest of the forms more easily. I always poke the looser concrete with a 2x4 to mix it with the stiffer concrete to eliminate cold joints, those distinct lines with small voids that can show up between the lifts. When I switched to a fresh truck near the other end, I again poured the stiffest mix into the forms where the addition foundation returned to the existing foundation. As with a regular foundation, I bring the concrete up to grade and float it flat and even with a 2x4 or a trowel. Then I place the anchor bolts.

The floor of the excavation for this addition was quite a distance below the existing foundation, so after stripping the forms, I set up a single-face wall to shore up the excavation (sidebar, p. 11).

Rick Arnold is a builder and residential-construction consultant in North Kingstown, Rhode Island. He is the author of Concrete Foundations *and co-author of* Precision Framing, *both published by The Taunton Press, Inc.*

Supporting an Addition

▪ BY PHILIP S. WENZ

A homeowner recently approached me with an idea for adding on to his house. His options were limited by his tiny lot, but he thought that adding a second floor to his house would afford him an excellent view of San Francisco Bay. I wanted to offer him words of encouragement, but while he was dreaming of the view from the second floor, I was worrying about whether his house's aging foundation and the building's marginal structure would be strong enough to support a second-story addition.

The Foundation Is a Critical Part of an Addition

If you're planning to add a second story, it's wise to assume that your foundation will not work unless it meets certain criteria. If a foundation is not sized properly as specified by the building code, it may not support the load of an added story. A foundation also needs to be solid, with no cracks or crumbling that could indicate structural weakness, again reducing capacity to carry additional weight. Finally, a foundation also should be reinforced with steel rod, called rebar. If it is lacking in any of these areas, a structural

Before adding on, check the structure. This house's siding and sheathing have been stripped, leaving the framework and foundation exposed for inspection and modification, if necessary.

engineer should be called in to assess the foundation. The foundation probably will need reinforcement that can vary from simple footings strategically placed beneath sections where the new second-story loads will be concentrated to complete foundation replacement.

Concrete technology has been evolving for a while, so older concrete foundations might not be as strong as today's. In the past builders occasionally added untreated lime and other contaminants to their mixes. Excessive water and aggregates of the wrong size and material also contributed to weakness in concrete. Foundations often were poured in freezing or extremely hot weather, which interfered with curing. Concrete with these problems eventually weakens and has left many of today's older foundations on the verge of failure. In the most extreme cases a structural engineer might suggest temporarily supporting the entire house while a new foundation is poured with the proper reinforcement prior to adding a second floor.

Adding Rebar Offers Additional Support

Rebar was seldom used in older foundations. Unreinforced concrete is strong under compression, but it has little tensile strength, which is not a problem as long as the foundation is loaded evenly and is supported equally by the soil beneath it. However, if the loading is not even or if the soil does not have uniform bearing capacity, the foundation will tend to bend between applied loads, or some parts of the house will settle at different rates. These conditions, known as differential loading and differential settling, place the bending parts of the foundation under tension. An unreinforced foundation will respond by cracking where it would otherwise bend if it had tensile strength.

Adding rebar makes the foundation act like a wood or steel beam that resists forces in both tension and compression. If a nonuniform load is applied at any point along the foundation wall, it will bend slightly and transfer the load along its entire length. Remember, however, that even a new, reinforced foundation with high-strength concrete may not be properly designed or sized for a second-story addition.

The concrete in some older foundations is simply too weak or damaged to be reinforced, and building a second story would require its complete replacement at a prohibitive expense. Another solution, however, is supporting the weight of a second floor on separate piers outside of the existing foundation (drawing, below). This system is called bridging and may eliminate the need for expensive rehabilitation of the existing foundation.

Bridging the Foundation

If an existing foundation won't support the weight of a second-floor addition, the addition can be supported on columns and piers outside the foundation perimeter, eliminating the need for modifying the existing foundation.

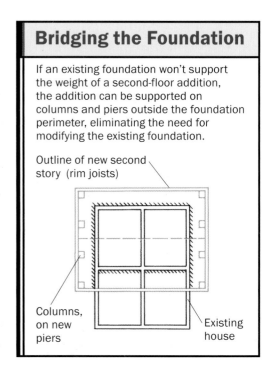

Outline of new second story (rim joists)

Columns, on new piers

Existing house

Foundation Problems Are Often Linked to Soil Conditions

The condition of the soil is as important as that of the foundation when planning an addition. The quality of the soil is equally important. For example, loose soil is a concern because it offers insufficient compressive resistance to the weight of the house. The differential settling that can result from loose soil can cause floors to tilt; walls, foundations, or chimneys to crack; and windows and doors to stick. A house built on loose soil probably will make a poor candidate for a second-story addition without remedial foundation work. Similarly, a room addition built beside a house and atop loose soil is likely to settle at a different rate than the existing house, creating extra stress where the two structures are joined.

Loose soil can exist in varying degrees. If your house shows signs of differential settling, a soils engineer should be called in to evaluate the site and offer advice. When you're building a room addition on a site with loose soil, a common recommendation is underpinning part of the old foundation with an extended piece of the new foundation where the two are joined (top drawing). This foundation design can help address the problem of differential settling between the two structures. If you plan a room addition, use rebar to connect the two foundations, regardless of the soil conditions (bottom drawing).

Another solution can be putting the addition on a pier-and-grade-beam foundation. This type of foundation consists of poured reinforced-concrete piers that go through the loose soil to bear on solid ground or support the vertical load by the friction of the earth on their surfaces. These piers support a reinforced-concrete beam poured on grade that the house or addition rests on. In addition to recommending this type of foundation for your addition, a soils engineer also might suggest integrating piers with your

Underpinning the Old Foundation with the New

One way of connecting the old and new foundations is by pouring part of the new foundation underneath the old. This procedure interlocks the two foundations, allowing them to function as one, thereby counteracting the effects of differential settling.

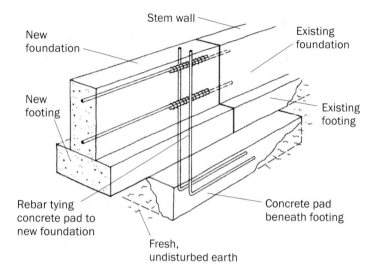

New foundation · Stem wall · Existing foundation · New footing · Existing footing · Rebar tying concrete pad to new foundation · Fresh, undisturbed earth · Concrete pad beneath footing

REBAR FOR ADDITION BEGINS IN EXISTING FOUNDATION

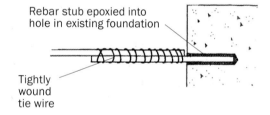

Rebar stub epoxied into hole in existing foundation · Tightly wound tie wire

Connecting Old and New Foundations

To attach the foundation of an addition to the original foundation, rebar is first drilled and epoxied into the original foundation. Additional rebar for the new foundation is wired to the embedded rebar to connect both footings and foundations.

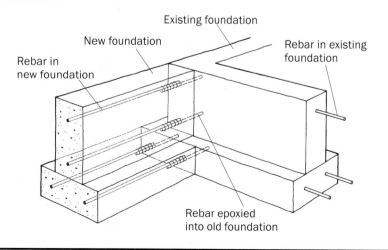

Existing foundation · New foundation · Rebar in new foundation · Rebar in existing foundation · Rebar epoxied into old foundation

existing foundation so that the foundations of both the house and the addition will behave similarly.

Uncompacted fill, where the earth on a site has been placed and not compacted by machine to the point at which it can support a building, can cause the same kind of settling problems that occur with naturally loose soil, but even more so. That's because uncompacted fill is often more profoundly disturbed than naturally occurring loose soil and may vary in density throughout the site. Be on the lookout for sunken areas or holes on your site that may indicate remnants of excavations, old wells, cisterns, or septic systems that have been filled or covered over. These areas indicate possible loose-soil conditions that again will not support the foundation of an addition. If you find any of these symptoms of loose soil, have the soil tested by an engineering firm before proceeding with the design.

Cracks in Retaining Walls, Driveways, and Foundations Indicate Soil Motion

The unsettled soils in many newer geological regions—including much of the West Coast, parts of the Rockies, and elsewhere throughout the country—tend to creep downhill. Technically, this movement is a very slow landslide. Dramatic, destructive landslides are rare. But building sites with creeping soil are common, and hundreds of thousands of houses are built on them. Creeping soil damages roads, sidewalks, and foundations as it moves irrevocably down the slope.

The problem with adding on to a house on a site where the soil is moving is making sure that the house and the addition move in the same direction and at the same rate. Your engineer will have to design a foundation system for your addition that will work in harmony with your existing foundation.

Foundation cracks are clues in a drainage puzzle. The cracks in this foundation are due to a combination of loose soil plus localized subsidence from the half-hearted attempt to divert roof runoff away from the house. Before an addition is built, new reinforced footings should be put under the foundation, and the poor drainage system needs to be corrected.

Cracks appearing all over the site are an indication of major soil motion.

When only a few cracks appear in a small area of the foundation but the rest of it is in good condition, localized soil subsidence is most likely the problem (photo, above). An example of this condition is a foundation's undermining and subsequent settling that occurs because of improperly diverted water from a downspout. The soil beneath the foundation becomes saturated and soft in these areas, allowing differential settling to occur.

Another source of trouble is expansive soil, which includes many types of clays found all over the country. Expansive soil swells as it absorbs water, creating hydrostatic pressure that can cause the foundation to heave and crack. An existing foundation that has been weakened by expansive soils will need to be repaired before an addition is built. An addition's foundation built in expansive soil may react differently than the house foundation, causing a variety of problems for both structures. The detrimental effects of expansive soils can be controlled to some extent by installing proper drainage.

A Good Drainage System Helps Keep the Site Stable

Inside the basement, efflorescence can be an indication of an outside drainage problem. Efflorescence appears on foundation walls as a white, crystalline deposit that is the result of water from around the foundation passing through concrete and leaching salt and minerals from within. These telltale deposits are left as the water evaporates on the inside surface of the concrete. Building an addition is a great opportunity to improve an undersize or deteriorating drainage system. The first concern is to avoid making the drainage situation worse by building a foundation that traps water, such as an L-shaped foundation with the inside of the L facing uphill. Driveways and the soil around the house also need to be graded to shed water away from the building, and the drainage system should discharge in a location where the diverted water won't pose problems for the house and addition.

If you have a drainage system on site but there is evidence of drainage problems, you need to figure out why the system has failed. Look first at those factors that are readily visible aboveground. Check for runoff from neighboring properties that might be collecting on your site. Then examine the condition of the gutters and downspouts. Downspouts on the low side of your house should discharge onto splash blocks, and high-side downspouts should be hooked to underground pipes that return to daylight a safe distance from the house and addition. You can test the underground pipes by sticking a garden hose in the downspout opening and turning it on full blast. If the flow from the pipe is sluggish or if the water shows up somewhere other than the end of the drainpipe, the system is clogged and should be cleaned out or replaced before an addition is built.

After the Foundation and Soil, Assess the House Framing

If your house has a wood frame and is more than 60 years old (like the house of my client who wanted to build a second story by the bay), chances are it is a balloon frame. These frames are characterized by their long, continuous studs, running from the mudsill to the second-story or even third-story plate (drawing, p. 18). The floors and internal walls were added after the shell was framed. Joists in a balloon frame typically are supported midspan by a girder. On the outside walls, the joists are nailed to the studs, and in most balloon frames a specialized ledger called a let-in ribbon provides additional joist support along the outside walls.

In some of the earlier balloon frames, these ribbons either were not let in or were omitted (top left drawing, p. 19). This situation left a floor supported at the walls by a handful of nails. In many older balloon-framed houses, the joists are undersize. This condition causes floors to sag and pull away from their exterior-wall supports. If your house is balloon framed, you need to examine these floor-framing details carefully before designing an addition. In most cases you will need to beef up the floor system before loading it with any additional weight from a second floor. A structural engineer's advice will be invaluable in determining the correct path of action. If your balloon-framed house lacks the support of a let-in ribbon, you may be required to add blocking under the joists and a ledger nailed to the studs to carry the weight of the floor. Your engineer also might recommend adding extra walls or girders to pick up the load of undersize or overspanned joists.

TIP

Before undertaking any addition project, assess the drainage situation. Symptoms of poor drainage include cracks in retaining walls and foundations, soil erosion, basement dampness, or mud in the crawlspace.

Assessing the House Framing

PLATFORM FRAME
With the advent of plywood and the scarcity of framing lumber for balloon frames, the platform frame became the most widely used method of house building. This type of framing is the most adaptable for attaching an addition.

BALLOON FRAME
The balloon frame is characterized by continuous studs running from the mudsill to the top plate. Floor joists are attached to the studs and rest on let-in ribbons.

Rafter
Ceiling joist
Rim joist
Double top plate
Stud
Plate
Double top plate
Floor joists
Stud
Rim joist
Floor joists
Sole plate
Plywood or board sheathing
Mudsill
Concrete foundation

Rafter
Ceiling joist
Floor joists
Double top plate
Fire blocking
Stud
Let-in ribbon
Fire blocking
Continuous stud
Let-in ribbon

Another problem frequently found in balloon-framed houses is undersize door and window headers. The weight of an added second story bearing on an undersize header will cause it to bend excessively, creating a variety of problems for the wall above and for the door or window below. The headers in some of the oldest houses were let into or simply nailed onto king posts or studs without trimmers (bottom right drawing, facing page). Under the load of a second-story addition, the nails holding these unsupported headers are liable to fail in shear. Before a

second-story addition is designed, these framing details need to be examined carefully and additional framing added to support the headers, if necessary.

If your house was built after the 1930s, it is probably platform framed (drawing, above). In a platform-framed house, the studs are not continuous from foundation to roof. The joists rest directly on the plates and usually span about half of the house with the other end resting on a central girder. Overspanned or unsupported joists are not as common a problem in platform frames.

Balloon Framing: Good, Bad, and Upgraded

Floor joists in a balloon frame are normally supported by ribbons, boards that are let in to the studs (1). Occasionally, the floor joists were just nailed to the studs, leaving the weight of the entire floor supported by just a handful of nails (2). To remedy this problem before building an addition, blocking can be added beneath the joists and a ledger nailed or screwed to the studs and blocking (3).

1. GOOD

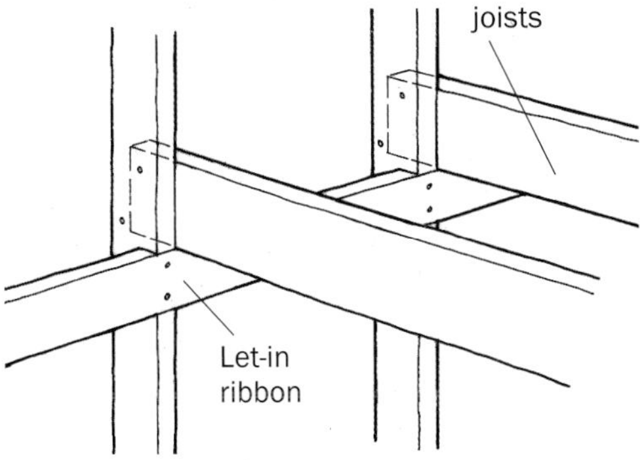

Floor joists

Let-in ribbon

2. BAD

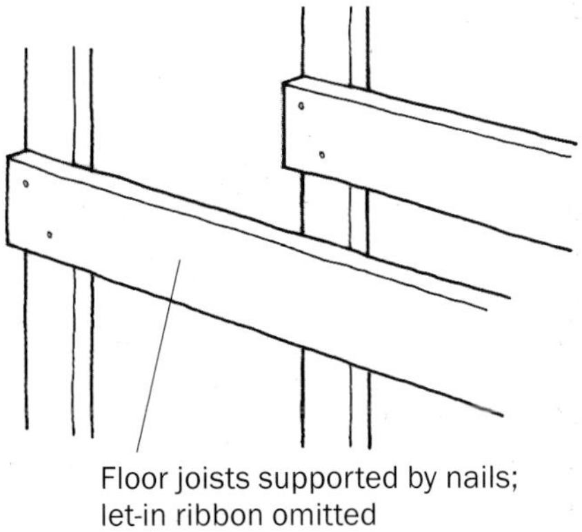

Floor joists supported by nails; let-in ribbon omitted

3. UPGRADED

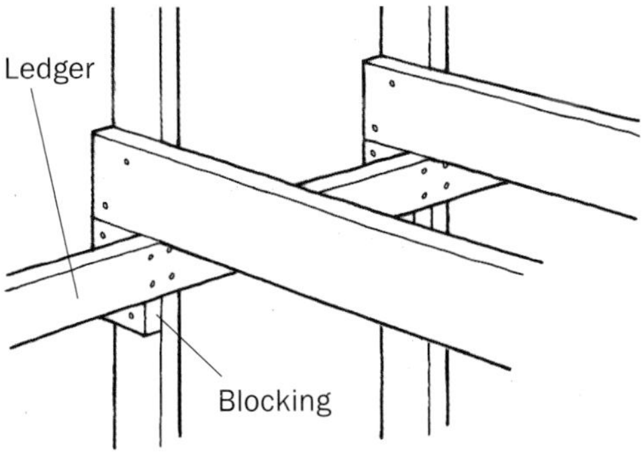

Ledger

Blocking

Blocking and ledger support the floor joists.

UNSUPPORTED HEADERS
In older buildings headers were often installed without supporting framework. Trimmers or jacks need to be installed before a second floor is added.

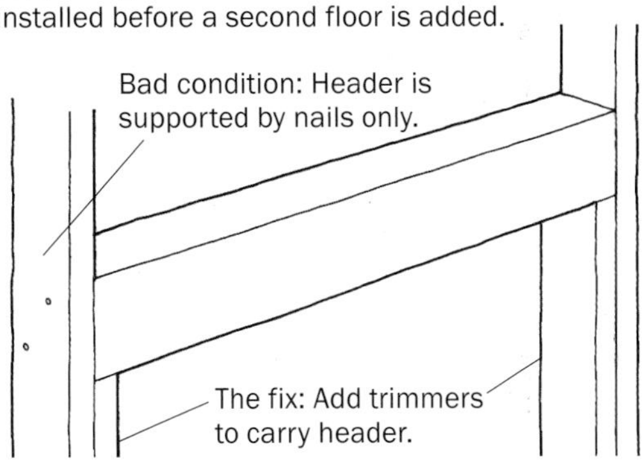

Bad condition: Header is supported by nails only.

The fix: Add trimmers to carry header.

The walls are usually sheathed with plywood and built in small sections, each with its own bottom and top plates. Plywood sheathing also lends a great deal of shear strength to the walls, helping them resist lateral forces generated by high winds and earthquakes.

When designing your addition, remember that the foundation, soil, and framing details must be carefully examined. Problems are often interrelated, and a set of symptoms may have more than one cause. Foundation cracks, for example, could develop because an unreinforced foundation with poorly mixed concrete is sitting on soil of varying density and is not uniformly loaded. The solution you choose must address all of the problems that exist.

Philip (Skip) Wenz is a designer who teaches building technology at the San Francisco Institute of Architecture. He is also the author of Adding On to a House, *published by The Taunton Press, Inc.*

Laying Up Concrete Block

■ BY JOHN CARROLL

Masons work from the ground up. You're probably not surprised to hear that. As simple as this principle sounds, however, it's also a bit misleading. Masons build from the ground up, but they measure from the top down, which means the real starting point of a masonry job is the top line.

From this line, masons measure down, marking in equal increments what will be the top of each course of brick or block. By working to these marks, masons arrive with evenly spaced courses at the tops of foundations, the bottoms of windowsills, or other predetermined landing places.

Even when I'm using basic techniques for planning and laying up a simple concrete-block foundation, I return again and again to the same starting point: the top line.

Planning and Laying Out a Block Structure

Before any mud is slung or any blocks are buttered—indeed, before any dirt is dug or any concrete footings are poured—the finished dimensions of a block structure should be established. These dimensions are length and width—along with any variations in the basic rectangle—and exact height.

When I build a foundation, for instance, the first thing I do is see if any small dimension adjustments might allow the masonry units to fit without being cut. Say this foundation is 12 ft. by 13 ft. I try to change the 13-ft. dimension to 13 ft. 4 in. so that it works out to an even number of blocks.

Unlike the foundation for a freestanding building, where a difference of an inch or two in the final height usually isn't too important, the height of the foundation for an addition must be right on the money. To get to this point, the first thing I do is to mark the height of the existing finished floor on the outside wall of the house. For this, I set up my laser level, shoot the elevation of the floor through an open door or window, then transfer that elevation to the outside wall.

Next, I find out exactly what will be used in the addition as floor covering, subfloor, joists, and sills. On the foundation project shown in this article, these included carpet and pad (¾ in.), tongue-and-groove plywood (¾ in.), 2x10 joists (9¼ in.), and 2x8 sills (1½ in.). By measuring down a total of these depths—12¼ in.—from the mark I made on the outside wall, I arrived at the correct

Using the proper layout, mortar consistency, and troweling techniques will ensure evenly spaced courses and consistent mortar joints.

height for the top of the foundation. Using my laser level, I made marks at this elevation at both ends of the planned addition, and I snapped a line on the house to represent the top of the new foundation.

Establishing the Entire Top-of-Foundation Line

The project pictured in this article is a 16-ft. by 24-ft. foundation for an addition. Here's how it began. After I snapped a line on the house to represent the top of the foundation, I marked the beginning and end of the foundation on this chalkline, which served as a reference to lay out the rest of the foundation.

My next step was to set up a line that would represent the top of the outside foundation wall. This line had to be 16 ft. from—and parallel to—the house and at the same level as the chalkline. To hold the string, I set up a pair of batter boards. First, I drove a pair of 2x4 stakes into the ground several feet outside the corners of the planned foundation (I could judge this by eye by looking at the corner marks on the chalkline marked on the house).

I set the stakes in line, 14 ft. and 18 ft. out from the house so that they would straddle the planned 16-ft. line. After driving in the 2x4s, I used my laser level to transfer the top-of-foundation elevation, represented by the chalkline, to the stakes. Then I attached a horizontal batter board to each pair of stakes, keeping the top edge even with the top-of-foundation marks. Once the batter boards were attached, I stretched a line from one batter board to the other. To hold the line in place, I used mason's line blocks, which I could slide along the batter board. By methodically measuring and adjusting the work I was doing, I was able to get the line exactly parallel to and 16 ft.

away from the house. This string represented the top outside edge of the foundation wall.

I now turned to the task of laying out the two sidewalls. These walls would run perpendicular to the house and would begin at the corners marked on the chalkline. To do this, I calculated the hypotenuse of a right triangle with sides of 16 ft. and 24 ft. This works out to 28 ft. 10⅛ in. I pulled this dimension diagonally across the planned foundation from one of the corner marks on the house to the string. Using a felt-tip pen, I marked where 28 ft. 10⅛ in. intersected the string. This dot represented the third corner of the foundation. Pulling this dimension diagonally in the other direction, I marked the fourth corner. To check my work, I measured from dot to dot along the string. Seeing that it was exactly 24 ft., I knew my layout was correct.

After the third and fourth corners were marked on the string, I set up two more pairs of batter boards. To represent the top of each wall, I stretched a string from the corner mark on the house, through the mark on the string, to the new batter board. I now had strings outlining the top outside edge of the entire foundation.

A Story Pole Aligns the Courses and the Corners

The top of the footing should be a set number of block courses below the top-of-foundation line. In the example that I'm using, the footing was exactly 80 in., or ten block courses, below the top line.

After the footings are poured, it's good to recheck the position of the top-of-foundation lines. After seeing that my top-of-foundation line hadn't moved, I dropped a plumb bob from the intersection of the lines to mark the outside corners of the block courses. At each outside corner, I attached a story pole

of 2-in. tubular steel to the footing. The story poles I use have two L-shaped brackets welded to their bottoms. To attach the story poles, I nail through holes in the brackets with case-hardened, fluted masonry nails into the concrete footing. Then I set the story pole plumb, and I clamped 1x4 braces to the pole and to the batter boards (photo, above).

At the corners that engage the house, I affixed 1x4 story poles. Next, I marked down all the story poles in 8-in. increments from the top-of-foundation line and was ready to lay block.

For Efficiency and Looks, Avoid Cut Blocks

Blocks are difficult and expensive to cut, and when they are cut, they can detract from the appearance of the job. So it's wise to avoid unnecessary cuts. Cutting units lengthwise (the masonry equivalent of ripping) is referred to as splitting. Cutting units to length (the masonry equivalent of crosscutting) is simply called cutting. Split blocks

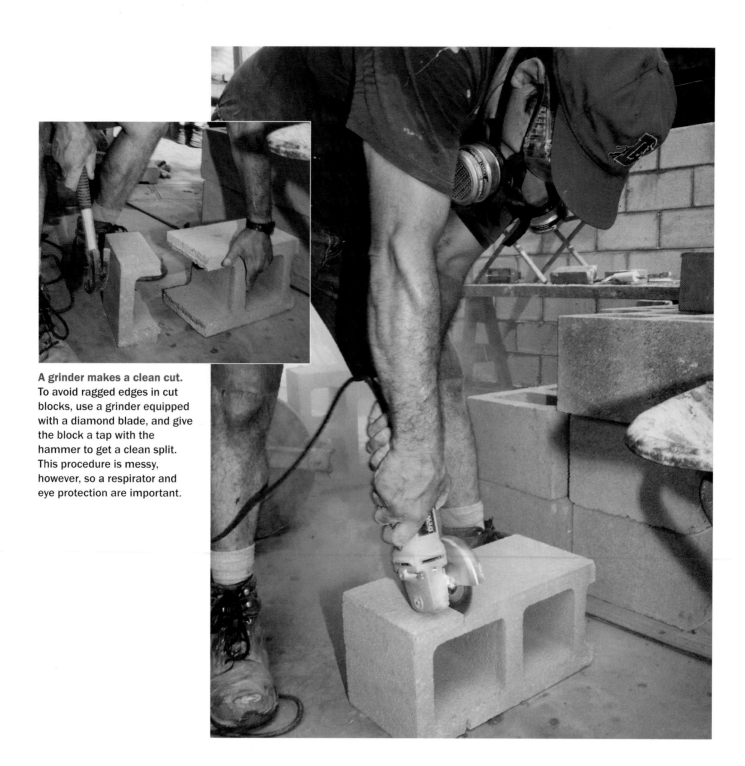

A grinder makes a clean cut. To avoid ragged edges in cut blocks, use a grinder equipped with a diamond blade, and give the block a tap with the hammer to get a clean split. This procedure is messy, however, so a respirator and eye protection are important.

not only interrupt the orderly progression of horizontal courses—and look bad—but they're also a scourge of productivity. Not surprisingly, masons do their utmost to avoid split units. Units cut to length are a different matter. Because every house has windows, doors, and corners, blocks inevitably have to be cut to length.

When I have to cut block, I avoid masonry hammers or chisels. I've never liked the ragged edges these tools leave. Instead, I use a 4-in. grinder equipped with a diamond blade (right photo). You can also use an abrasive blade in a circular saw. After cutting as far into the block as I can, I give it a good tap with a hammer, and it breaks easily (left photo).

Buckets Are Better Than Shovels when Mixing Mortar

Like all troweling tasks, block work is mainly a physical skill that takes a lot of practice. The best training is to have at it, ideally alongside an experienced mason.

The first thing a novice should learn is how to make good, consistent mortar. I've found that this cannot be done using ready-mix mortar. The material that comes already mixed with sand has poor plasticity and contains coarse sand.

For mixing mortar, you can rent electric mortar mixers for between $30 and $50 a day. However, most novices are laying block so slowly that they'd be just as well off mixing it in a wheelbarrow by hand. They're not using mortar fast enough to justify the cost of renting a machine.

Mortar is made up of portland cement mixed either with lime or any of several proprietary ingredients. One part of this mixture is combined with 2½ parts to 3 parts sand, and water is added to get the right consistency. Cement/lime mortars often come in separate bags and have to be mixed on the job. Masonry-cement mortars (those with the proprietary ingredients) come ready to be mixed with sand.

Masons often count 18 shovels of sand for every bag of mortar, but the size of a shovel is inexact. I measure the ingredients by filling a drywall bucket with dry mortar and three other drywall buckets with sand. I put all this in a mixer and add water until I get the right consistency.

It's hard to describe the right consistency. As opposed to concrete, which should be kept as stiff as possible, mortar should be made as wet as possible yet still be workable (photo, below). The primary role of mortar is to bond masonry units together. Wet mortar spread on dry units is absorbed deep into the pores and crevices of the units, producing a tenacious bond. A mixture that is too wet, however, is almost impossible to work with and makes a mess of the job. Good mortar is almost fluffy; some masons call mortar that's just right "fat mortar."

The best way to learn how to make good, wet, workable mortar is by actually making it and using it. Even perfect mortar doesn't stay that way for long. On hot days, you often have to "shake up" the mortar by mixing in a little water. Enthusiastic novices invariably mix too much mortar. The longer mortar sits, the harder it is to work. After two hours, it should be disposed of, usually into the cells of the block wall.

I try to make about an hour's worth of mortar at a time. Usually one bucket of mortar and three buckets of sand last me about an hour; however, a novice may want to start by mixing a half-batch. And before I mix the material, I make sure I have everything ready. To estimate what you'll need, figure three bags of mortar for every 100 blocks plus 9 cu. ft. of sand, or about 14 5-gal. buckets (a 5-gal. bucket is 0.668 cu. ft.).

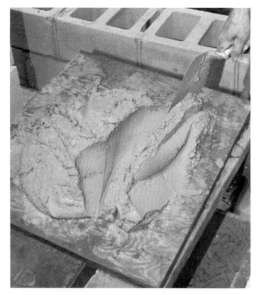

Not too mushy and not too stiff. Mortar should be as wet as possible yet still be workable. Mortar should have some body, but it still should be mushy.

Consistent Mortar Joints Testify to Good Work

Masons adjust the height or length of courses by altering the thickness of mortar joints. This is a basic part of masonry. It allows masons to make up for inconsistencies in the size of the units and to fit whole units into a given space (between windows, for instance). But this device is easily overused. Fat joints and abrupt changes in the thickness of joints look terrible.

Joints that are thicker than ¾ in. can shrink excessively, which sometimes results in leaky hairline cracks. And because masons tend to use a stiff mix when making thick joints, the bond is often poor, resulting in both leakage and structural compromise. For optimal appearance and performance, joints should be between 5⁄16 in. and ½ in. thick, and they should be as consistent as possible.

The height increments for concrete blocks are typically 8 in. or 4 in. Like lumber, blocks are smaller than their nominal size; an 8-in. block is actually 7⅝ in., and a 4-in. block is 3⅝ in. A block with a ⅜-in. bed joint, then, measures an even 8 in. or 4 in.

Getting Started with the First Course

After stretching stringlines for the first course in both directions from the corner pole, I lay my first block, a corner block. The top of this block fits snugly to the intersection formed by the strings.

After getting this block even with both strings, I start laying blocks in from the corner along one of the lines. After laying all or part of this line, I work down the other line, again working from the corner in.

After running all or part of this line, I move both lines up to the next mark on the story pole and start the second course. Like the first block I laid in the previous course, the first one here fits snugly into the intersection of the lines at the story pole. This block, however, crosses to the corner block in the first course and thus begins the familiar bond pattern associated with most block walls.

At this point in the job, I have the option of either laying the entire length of each course or laying just far enough down the line to build up the corner. Usually, there's no need to build a corner lead. I often build entire small foundations straight off the story poles. On congested sites, however, it's often best to build the corner first so that you can get the story pole and braces out of the way as soon as possible. To set up a line for laying the rest of the wall, just hook the line block directly to the corner you've built.

The ability to set up a line quickly and securely is an important masonry skill. The two most common tools for doing this are line blocks and twigs (photo, facing page), which masonry-supply houses traditionally supply for free. To use a line block, wrap the string over and around the block a couple of times and hook it on either the outside corner of the masonry or on a corner pole.

Twigs fasten to the string, and a brick or a piece of block holds it on the top edge of the block. One of the advantages of using a twig is that the string can be pulled taut without exerting pressure on the masonry just laid. Sometimes, instead of hooking a line block on a corner I've laid, I attach the line to a stake or a block beyond the corner and use a twig to hold the line even with the top edge of the unit.

A third tool that is sometimes used to affix lines to the inside corners is the line stretcher. After the corner has been built, the string is wrapped around the line stretcher, and the stretcher is placed across the top of the first unit of the course being laid. The stretcher is held in place by tension on the line. All these tools are available from a good masonry-supply company.

While you're pulling together the tools and materials you need to get started, you should also consider lightweight block

The basic tools for working with concrete block. To get started laying up block requires a trowel, a pointing tool, line blocks, and string. Also helpful are line stretchers, left, and "twigs," shown next to the string.

instead of standard block. Lightweight block is made with special lightweight aggregates rather than crushed stone. A standard 8-in. by 8-in. by 16-in. block weighs 32 lb. to 38 lb. A lightweight block weighs only 22 lb. to 27 lb., but it's weight that really adds up after a day of lifting. Lightweight block also costs up to 25% more than normal-weight block.

Troweling Techniques: To Sling or to Butter

There are two techniques for spreading the bed joints, or horizontal joints, of a block wall. In the first, the mason loads his trowel, holds it above the top edge of the lower block, and shakes some mortar loose from the trowel. Then, as he moves the trowel down the line toward himself, he shakes more of the mortar loose and lets it fall in a line along the block. After about a third of

the mortar is laid down like this, he simultaneously turns his wrist downward and pulls the trowel quickly toward himself, slinging the rest of the mortar in a nice line on top of the block.

In the second technique, the mason loads his trowel and gives it a good shake (a hard shake makes the mortar stick to the trowel). Then, in a downward, pulling stroke, he butters the top edge of the block by sliding the trowel blade down and across the block.

There are also two methods for buttering the head joints of blocks. Some masons butter the ends of the block before they pick it up and set it in the wall. Others butter the ends of the block after they set it in the wall. Either way, you need to apply it with some oomph so that the mortar hits the surface hard and sticks.

Finally, a few words on the trowel, which is the most basic tool in masonry. For me, the size of the trowel is analogous to the size of a hammer. Some carpenters use a 24-oz.

TIP

To attain evenly spaced courses with consistent mortar joints, masons lay out courses in standard modules based on the size of the block used.

hammer, and some masons use a 13-in. trowel. A small trowel is 10½ in., and a large trowel is 13 in. My suggestion is to use whatever is most comfortable in your hand. I use a small hammer and a small trowel because I have tendinitis. So whatever you use depends on how much weight you want to handle.

Getting the Rhythm of Working with Mortar and Block

Once I get going, the basic rhythm of masonry is to spread a bed of mortar an inch or so high, then gently push or tap the block until it's even with the line.

I like to keep about a ¹⁄₁₆-in. space between the line and the block because if the block touches the line, it can push it out (photo, below) and throw off a course. I almost always set the line up on the far side of the wall so that I don't have to lift blocks over it all day. As I set each block in the wall, I use

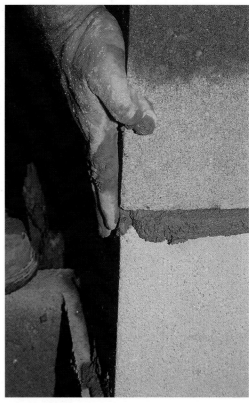

You can feel if a block is aligned with the rest. Place the palm of your hand on the new block and your fingers across the joint between blocks to feel if the new block is even.

Tap the block gently to get it plumb and level. This block was set into about an inch of mortar and gently tapped with the butt of the trowel until it was level and plumb. The layout line remains a fraction of an inch above the block to keep the string from snagging on the block.

the heel of my left hand (I'm right-handed)
to feel when the top of the block is even
with the preceding block (top photo, facing
page). I get this corner even by feel and,
looking straight down over the line, push or
tap the outside of the block even with both
the string and the block below it. At the
same time, I push the block horizontally
against the buttered ends of the preceding
block until the joint size looks right.

Using the trowel in my right hand, I cut
the mortar that has bulged out from the
joints (photo, above) and, giving my trowel
a good shake, butter the end of the block
I just laid (right photo). I do any minor
adjustments immediately. After blocks have
set for a while, they should not be disturbed.
Tapping blocks laterally after the initial set
of the mortar breaks the bond and weakens
the wall.

Clean the joint of excess mortar. After a block is set in place, use the trowel to cut the mortar that has bulged out. Then give the trowel a shake and butter the end of the block with the remaining mortar.

Tooling the joints for a neat look. When the mortar is as stiff as putty and beginning to pull away slightly from the edges of the blocks, it's ready for tooling.

Another important consideration in block work is keeping the blocks dry. Although dry and absorbent brick sometimes needs to be dampened before it is laid, blocks should always be kept dry before they are installed. At the end of the day, it's also smart to cover the top of the wall, both to protect the bond of the day's work and to make sure the top course is dry when you return to lay blocks on top of it.

To finish off the wall of the foundation, I install anchor bolts at every corner and at each door and window jamb, as the local code requires. I also installed bolts at least every 6 ft. between these points. To anchor the sills, I drop broken bits of block or brick into the cores to provide support for the grout, which I make out of concrete or a portland cement-sand mix, which is stronger than mortar.

As for tooling the joints, it's necessary to wait until the mortar is thumbprint hard before tooling begins. When the joints are ready, they are usually about as stiff as putty and starting to pull away slightly from the edges of the blocks. Using a jointing tool, I tool the vertical joints first, then do the horizontal joints (photo, above). Wherever the joints need a bit of mortar, I push a lump into the joint with my jointer.

There are many different jointers on the market. I own several different types. But for laying block, the S-shape jointer is a standard. You don't get into much variation in jointers until you get into brickwork. The S-shape jointer has two different-size jointers, with one at each end. Use whichever end looks best. I usually use the widest one so that it doesn't cut the joint too deeply.

TIP

Wet blocks don't bond well and are difficult to lay, so it's prudent to cover blocks as soon as they're delivered.

"Ladder" reinforcement equals lateral strength. Ladder-type wire reinforcement can be added to the bed joints of a block wall as it's laid up. In this example, the ladder reinforcements were added at every other course.

Sometimes, a Block Wall Requires Reinforcement

There are several ways to reinforce a unit masonry wall. The wall can be thickened by switching from 8-in. to 10-in. or 12-in. blocks. Ladder-type wire reinforcement can be added in the bed joints as the blocks are laid up (photo, above). You can apply stucco reinforced with fibers to the outside of the wall after it's built. Or you can reinforce the wall with rebar and concrete grout.

To design a reinforced-block wall, it's best to hire a structural engineer for the project.

You can also get an idea of what works in the area where you live by talking to other builders and to your local building inspector. Also, you can see what doesn't work by carefully looking in your area at foundations and retaining walls that have failed.

John Carroll is a builder and mason in Durham, North Carolina. He is the author of Measuring, Marking, and Layout: A Guide for Builders *and* Working Alone, *both published by The Taunton Press, Inc.*

A Builder's Screen Porch

■ BY SCOTT McBRIDE

My grandfather lived alone in a little bungalow by the seashore. We got to know each other in his final years by spending long summer evenings out on the screen porch. We talked about the many things the old man had done in his life and some of the things a young man might do with his. Sometimes we didn't talk at all—just listened to the waves and the pinging of the June bugs off the screen, watched the lights, smelled the breeze.

A screen porch at night can have a magic all its own, balancing as it does on the cusp between interior and exterior space. A porch offers just enough protection from the elements to foster relaxation and reflection, without shutting out the sounds and the smells of the cosmos. This dual nature of screen porches can make them difficult to build with style because the usual rules of interior and exterior construction often overlap in their design.

When the time came to build a screen porch on my own house here in Virginia, I had the luxury of time—no anxious client, no deadline, and no hourly wages to worry about. So I included lots of special details that I hope will spare my porch some of the problems I've seen in 20 years of remodeling other people's houses.

The Foundation

I sited my screen porch two risers up from grade and three risers down from the adjacent kitchen. This made a smooth transition to the yard without requiring too much of a descent when carrying an armful of dinner plates from the kitchen. To anchor the structure visually, I ran a continuous step of pressure-treated lumber around the perimeter as a sort of plinth (photo, below).

The step is supported by pressure-treated lookouts that cantilever off the poured-

A single pressure-treated step runs all the way around the outside of the porch as a sort of plinth.

The screen porch the author built on his own house combines Victorian detailing with a builder's considered construction methods.

concrete foundation (photo, below). I used pressure-treated 2x8s for the lookouts, inserted them into my formwork, and actually poured the concrete around and over them. There isn't much concrete above the lookouts, so to key each lookout into the mix, I nailed a joist hanger on both sides. A week after the pour, the projecting lookouts were rock solid.

A Hip-Framed Floor

Masonry is the obvious choice for the floor of a screen porch because water blowing through the screens won't affect it. Also, in hot weather the coolness of a masonry floor feels good on your bare feet. On the down-side, masonry is, well, hard. It's also difficult to keep clean, it's gritty underfoot, and it retains moisture in damp weather.

Open decking is a good alternative to masonry, as long as it's screened underneath to keep the bugs out. Spaced, pressure-treated

Lookouts embedded in the concrete (and held securely by the addition of a joist hanger nailed to each side) provide rock-solid support for the first tread of the step that runs around the porch's perimeter.

A coping of pressure-treated 2x8s supports the porch posts. Weep channels in the coping and an aluminum pan divert rainwater blown through the screens.

This floor system, which is framed like a shallow hip roof, allows water to run off the porch floor. Strapped joists bring the finish floor flush with the 2x8 coping.

eventually decay while the wood stays sound just a foot or so in from the drip line of the eaves. By bordering my floor with a treated coping, the untreated yellow-pine flooring would be recessed further under cover. Also, the coping would allow me to lay the tongue-and-groove (T&G) floor at the end of the job because the structure above—the roof and its supporting columns—bears on the coping, not on the flooring. A temporary plywood floor endured weather and foot traffic during construction and allowed me easy access to run wires in the 1-ft. deep crawlspace.

To ensure positive drainage, and to avoid standing water on the T&G floor I had decided to use, I pitched the floor ¼ in. per ft. from its center in three directions. This meant that I'd have to frame the floor like a shallow hip roof (photo, left). What became the ridge of the floor framing was supported by concrete piers.

I ran 1x strapping perpendicular to the joists and eventually laid the flooring over the strapping. In addition to promoting good air circulation under the flooring, the strapping served two other purposes: It allowed the flooring to run parallel to the slope so that most of the water would flow by the joints in the flooring rather than into them. The strapping also brings the top of the 1x flooring flush with the 2x coping. I could have used pressure-treated 1x for the coping, but because the roof and its support-ing posts rest on the coping, I wanted it to be substantial.

The joint between the ends of the floor-ing and the inside edge of the coping gave me pause. I knew that wind-driven water was likely to seep in here and be sucked up by the end grain of the flooring, leading to decay. I thought about leaving the joint intentionally open, say ¼ in., but I knew that such a gap would collect dirt and be an avenue for critters. Instead, I back-cut the ends of the floorboards at a 45° angle and let them cantilever a couple of inches past the strapping for good air circulation under-

yellow pine will make a good, serviceable floor, and having a roof overhead will pro-tect the floor from the harsh sun that is the nemesis of pressure-treated lumber. But open decking looks utilitarian at best, and my wife and I wanted something a bit more refined.

I decided to use untreated kiln-dried yellow-pine flooring, bordered by a coping of treated 2x8 (top photo). I have repaired a lot of old porches, and I have noticed that it's the outer ends of the old floors that neath. Meanwhile, the long point of the mitered end butts tightly to the coping.

To collect any water that might seep through the joint, I formed aluminum pans that run underneath the coping and lip out over the floor framing (top photo, facing page). I cut weep channels in the underside of the coping with a dado head mounted on my radial-arm saw to let water out and air in. I have since heard that aluminum reacts with the copper in treated wood, so I probably should have used copper for the pans.

Hollow Posts and Beams

The roof of a screen porch is generally supported by posts and beams rather than by walls. Solid pressure-treated posts work well for support, but they won't accommodate wiring or light switches. Solid posts also are prone to shrinking, twisting, and checking.

I made hollow posts of clear fir, joining them with resorcinol glue. Biscuits provided registration during glue up (middle drawing). I rabbeted the sides of the posts to receive both the frames for the screen panels and the solid panels below the screens. The bottom of each post was rabbeted to house cast-aluminum post pedestals. The pedestals keep the bottoms of the posts dry. They also allow air to circulate inside the posts to dry up any internal condensation. Rabbeting the pedestals into the posts makes them almost invisible and ensures that all rainwater is carried safely down past the joint between the pedestal and the post.

Because the 2x8 coping on which the pedestals bear is pitched (because of the hipped floor framing), I used a stationary belt sander to grind the feet of the pedestals to match.

Inland Virginia where I live doesn't get the wind of the Florida coast, but we get plenty of gales, and last year a tornado ripped the roof off a Wal-Mart® in another part of the state. To provide uplift resistance for my porch roof, I bolted the tops and

Porch Posts: Construction and Attachment Details

To prevent uplift from strong winds, the hollow posts are bolted at the bottom to the 2x8 coping and at the top to the rough beam.

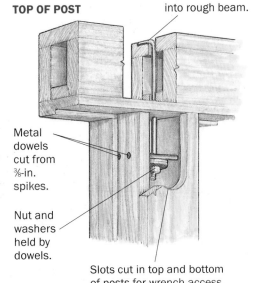

TOP OF POST

J-bolts mortised into rough beam.

Metal dowels cut from ⅜-in. spikes.

Nut and washers held by dowels.

Slots cut in top and bottom of posts for wrench access.

MIDDLE OF POST

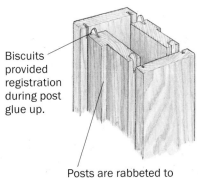

Biscuits provided registration during post glue up.

Posts are rabbeted to accept screen panels and solid panels.

BOTTOM OF POST

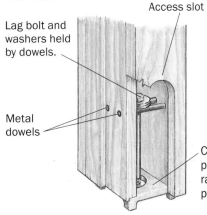

Access slot

Lag bolt and washers held by dowels.

Metal dowels

Cast-aluminum post pedestal rabbeted into post bottom.

Floor Framing and Post Details

The porch is supported by a series of hollow posts. Plywood wainscot panels provide lateral rigidity. The wainscot panels and the shop-made screen panels fit into the rabbets cut into the posts.

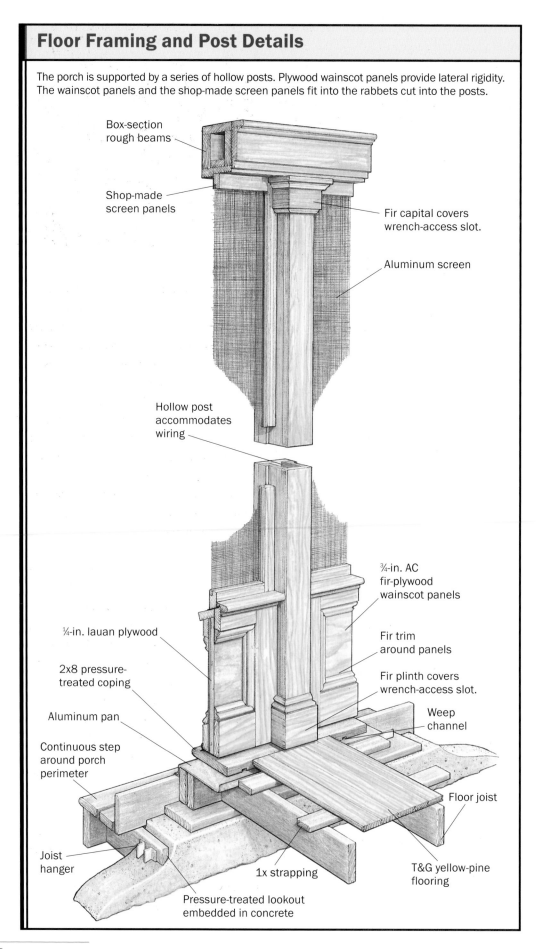

Box-section rough beams

Shop-made screen panels

Fir capital covers wrench-access slot.

Aluminum screen

Hollow post accommodates wiring

¾-in. AC fir-plywood wainscot panels

¼-in. lauan plywood

Fir trim around panels

2x8 pressure-treated coping

Fir plinth covers wrench-access slot.

Aluminum pan

Weep channel

Continuous step around porch perimeter

Floor joist

Joist hanger

1x strapping

T&G yellow-pine flooring

Pressure-treated lookout embedded in concrete

bottoms of the posts in place. Rather than relying on weak end grain to hold the bolts, I ran horizontal pairs of steel dowels through the posts, 3½ in. from the top and the bottom (top and bottom drawings, p. 35). The dowels were hacksawed from ⅜-in. dia.-spikes. At the bottom I passed a lag bolt vertically between the dowels and screwed it down into the floor framing until the head of the lag came to bear against the dowels (bottom drawing). At the top I used a similar arrangement, but instead of lag bolts, I used inverted J-bolts with the foot of the J mortised into the top of the rough beam, and the threaded end passing between the dowels. To get at the bolts with a wrench, I cut slots on the interior sides of the posts, which would be covered later with base and capital trim. I was surprised how rigid the posts felt after being bolted upright, even before they were tied together at the top.

The rough beams were made up with a box cross section rather than simply doubling up 2xs on edge (drawing, facing page). This gave the beam lateral as well as vertical strength so that any unresolved thrust loads from the untrussed secondary rafters above would be resisted by the horizontal top plate in the beam.

The Roof and the Ceiling

The inspiration for the coffered cathedral ceiling came from several sources. I once watched Japanese carpenters raise the frame of a small farmhouse. The delicate grid of the peeled white timbers against the sky made a lasting impression. I've also worked on Victorian houses in the Hudson Valley that featured finely wrought coffered ceilings over their verandas.

The framing scheme I finally decided upon is one that's found in some New England timber frames: trussed pairs of principle rafters interspersed with lighter, untrussed secondary rafters.

To create the curves on the bottom edge of the secondary rafters, the author first rough-cut the edges with a jigsaw, then trimmed them using a template and a router fitted with a flush-trim bit.

Instead of using heavy timber, I laminated each principle rafter in place from a 2x6 sandwiched between two 2x10s. Offsetting the bottom edge of the 2x6 helped disguise the joints, and the hollow channel above the 2x6 was useful for wiring.

Collar ties connecting principle rafter pairs have a 2x6 core sandwiched between 1x8s. The ¾-in. thickness of the 1x8 avoids an undesirable flush joint at the end where it butts into the rafter.

The secondary rafters are as wide as the principle rafters at the base, but their lower edges immediately arch up into a curve that reduces their width from 9 in. to 5 in. The constant width of all the rafters at the base allows the bird's mouth and frieze-block conditions to be uniform, even though the rafter width varies. I roughed out the curve of the secondary rafters with a jigsaw, then trimmed them with a flush-trim router bit guided by a template (photo, above).

Short 2x4 purlins span between the rafters on approximately 2-ft. centers (top photo, p. 38). The ends of the purlins are housed in shallow pockets routed into the rafters, also with the help of a plywood template. I fastened the purlins with long galvanized screws.

The roof-framing material was selected from common yellow-pine framing lumber. Before I remilled the lumber, I stickered it and covered it with plywood for two months to let it dry.

Primary and secondary rafters combined with a series of purlins comprise the porch's roof system. The secondary rafters curve along their bottom edges to reduce their width from 9 in. to 5 in. The purlins are let into the rafters and secured with screws.

The author used a chicken ladder—a narrow set of stairs built on site—to ease the task of installing the vertical sheathing that runs from the eaves to the ridge.

The roof was sheathed with 2x6 T&G yellow pine run vertically, perpendicular to the purlins. The exposed V-joint side faces down, and the flush side faces up. Running the boards vertically added to the illusion of the porch's interior height; it was a pain in the neck to install because I had to maneuver from the eaves to the ridge while nailing each piece. To facilitate the process, I built a chicken ladder—a narrow staircase that hooks over the ridge and runs down to the eaves (photo, above).

Building a structure with an exposed finished frame was difficult and time-consuming. Floor space in my shop was strained to the max while all the components were fabricated. Everything had to be given multiple coats of a water-repellent finish to prepare it for the eventuality of rain before I could dry in the structure—I used Olympic WaterGuard® from PPG Industries, Inc. Moving ladders and scaffolding around all that finished woodwork was harrowing. The payoff, though, was a structure with a kind of bare-bones integrity that would have been hard to achieve with the conventional approach of rough framing wrapped with finish material.

Finish Details

To contrast with the yellow pine in the ceiling and the floor, I used fir for all the woodwork from the floor up to the interior frieze. The choice of fir allowed me to order matching stock screen doors, and this saved a lot of time in the shop. To reinforce the doors against racking, I introduced slender diagonal compression braces into the doors' lower screen panels.

The structure itself gains much-needed shear strength from the wainscot below each screen panel. The wainscot has no interior framing: It is built up with plywood and trim boards. First I screwed ¾-in. AC fir plywood panels to the posts, good side in. I bedded the panels into the same rabbets that would receive the screen frames above the wainscot. I then attached 5/4 fir rails and stiles to the inside face of the fir plywood. To avoid exposed nail heads, I screwed through the back of the panel to catch the trim.

On the outside, I tacked a sheet of ¼-in. lauan over the back of the AC plywood. Lauan holds up well in exterior applications and takes a good paint finish. The stiles and the rails on the outside were nailed through both layers of plywood into the interior stiles and rails. The resulting sandwich proved remarkably stiff. I capped the panels with a beveled sill and a rabbeted stool.

For drainage, the bottom edge of the wainscot was raised 1 in. above the floor

Yellow pine and Douglas fir complement one another on the interior of the porch. The rafter system, the vertical roof sheathing, and the flooring are all yellow pine, while the posts and the panels are Douglas fir.

coping. To keep bugs out, I stapled a narrow skirt of insect screen around the outside. The top of this skirt was clamped down with a thin wooden band. A similar condition was achieved at the doors by attaching sweeps of insect screen. I even weather-stripped the edges of screen doors using a compressible-rubber weatherstripping.

When it came time to lay the T&G floor, I pondered the best way to deal with the shallow hips where the pitch of the floor changes direction. Rather than have a continuous 45° joint, which would be prone to opening up and collecting dirt, I decided to weave the floorboards in a herringbone pattern (photo, below). Working from the longest boards out to the shortest, I grooved the end of each board so that it would engage the leading tongued edge of its neighbor. To cut the end groove, I used a ¼-in. wing cutter chucked in a router (photo, facing page). The result is a pleasing stepped pattern that is accentuated by the way sunlight bounces off the wood according to the grain direction and the different planes of the hipped floor. Depending on where you stand, the floor has almost a faceted look; one side of the hip looks darker than the other.

Outside, I finished the porch with details consistent with my late 19th-century house.

I extended the cornice return all the way across the gable by cantilevering lookouts off the gable studding. This creates a full pediment and gives the porch's gable end the same overhang protection as its eaves. The tops of the posts sport scroll brackets on the outside and simple capitals on the inside.

Screen for the Porch

I made wood frames for my porch screens out of 1x2 fir. I used mortise-and-tenon joinery with an offset shoulder on the rails. The strength of a mortise-and-tenon joint isn't really necessary for a fixed frame that gets fully supported in a larger structure. But the design of a mortise-and-tenon joint makes it easy to use a table saw to cut the rabbets and plow the spline grooves before assembling the frame.

Spline stock holds the screen in the frame. Tubular in cross section, the spline stock gets pushed into a groove on the frame where its compression holds the screen in place. Spline stock is made from rubber or vinyl, and it's available in a smooth profile or with ridges around the circumference. The ridges help guide the splining tool, and they give the spline a little more bite on the walls of the groove.

The tool used to press in the spline looks like a double-ended pizza cutter. One disk

The hipped floor slopes in three directions to shed water that blows through the screens. Sun hitting the finished floor gives a pleasing effect. The joists are cross-strapped, and the flooring is laid on the strapping so that it runs parallel to the slope of the porch floor.

Drawings: Bob Goodfellow

Sources

PPG Industries, Inc.
1 PPG Place
Pittsburgh, PA 15272
412-434-3131
www.ppg.com

A router grooved the end of each piece of flooring so that it could herringbone its way down the floor's hips.

has a convex edge used initially to crease the screen into the groove. The other disk has a concave edge, which tracks on the round spline as it is pressed into the groove.

The two most common types of screen are aluminum and vinyl. Aluminum screen is available in mill finish or charcoal.

I used mill-finish aluminum for my screen porch because it seemed to be the most transparent. I also think aluminum is somewhat stronger than vinyl and less likely to sag over wide spans. The main drawback of aluminum is oxidation, which gradually forms a grainy deposit on the wire and reduces the screen's transparency. I live in a rural inland area where salt and pollution aren't prevalent. If I lived near the sea or in

an urban environment, I would have leaned toward vinyl. I would also go with vinyl if I were hanging the screen in place vertically, rather than rolling it out on a bench. Vinyl is much easier to work with and less likely to crease. A final consideration in choosing screen is the resounding ping made by bugs slamming into a tightly stretched aluminum screen. I rather enjoy it—it's one of the unique sounds of summer—but others might prefer to muffle the impact by using the softer vinyl screen.

Scott McBride is a contributing editor of Fine Homebuilding *and the author of* Build Like a Pro: Windows and Doors, *published by The Taunton Press, Inc. He lives in Sperryville, Virginia.*

A Dining Deck

■ BY TONY SIMMONDS

One of the joys of living in Vancouver, British Columbia, is that when the weather is warm enough, you can open your house to the outdoors without being eaten alive by insects. So it's a natural feature of local design to include a deck or some outdoor living space in your house plans. I enjoy the challenge of integrating outdoor spaces with the rooms that border them.

Bill Abbott and Kris Sivertz's house had a bad deck in a good location. Right off a small eating area in the kitchen, the deck was on the south side of the house between the kitchen and the backyard pool. That was the good news. Unfortunately, the deck was detailed poorly, and as a consequence it was falling apart from rot. It was also one step down from the floor level in the house. Besides being a potential hazard, this broke the continuity of floor surface, which is one of the essentials of good indoor-outdoor connections.

The other essential is a generous opening between the two, and the 5-ft. patio slider Bill and Kris had didn't qualify. Even the widest of sliders fails in this respect because half the opening always is obstructed. What you need is hinged French doors; by opening them up wide, you truly can abolish the barrier between inside and outside.

Here, though, we had a problem. There wasn't space on either side of the opening for a 2-ft. 6-in. door, let alone the 3-ft. doors I planned to use. They were going to have to be bifolds, an arrangement certain to cause some difficulties. I consulted Peter Fenger, custom sash and doorman, and he didn't say it couldn't be done, so we plunged ahead.

Benches and Planters Take the Place of Railings

To feel like an extension of the indoors, an outdoor space needs some enclosure. Usually, walls are suggested by a railing and the ceiling by a trellis. But the 42-in. guardrail required by code can make a small deck feel like a crib. Luck was on our side in this instance, however. Because the deck is less than 2 ft. above grade, we were free from the railing-code constraints. So I planned benches for two sides of the deck, enclosed and supported at their ends by planters (photo, p. 44).

Opening the doors doubles the room. A generous, 6-ft.-wide door opening coupled with matching floor levels creates an intimate connection between the garden deck and the dining room. Cedar trim inside and out intensifies the bond.

Partly in, partly out. Built-in benches and planters at the corners make a boundary around the sides of the deck, and the spare trellis overhead implies a ceiling. The planters brace the posts at the corners.

I believe strongly that an indoor/outdoor space needs at least the suggestion of a ceiling. So I wanted an open trellis of 4x4 beams above the deck. But I had to work hard to convince Bill and Kris of the wisdom of this idea. Given the scarcity of sun that shines here, people are reluctant to put anything between themselves and the sun.

Deck Details for a Wet Climate

Bill and I framed the deck with pressure-treated 2x8 hemlock joists, supported on a ledger at the wall and on a triple 2x8 built-up beam. Instead of spiking the 2x8s together, I used the code alternative of ½-in.-dia. bolts, 40 in. o.c. I used extra washers to space the 2x8s about ½ in. apart—enough to allow air

Section through Bench and Planter

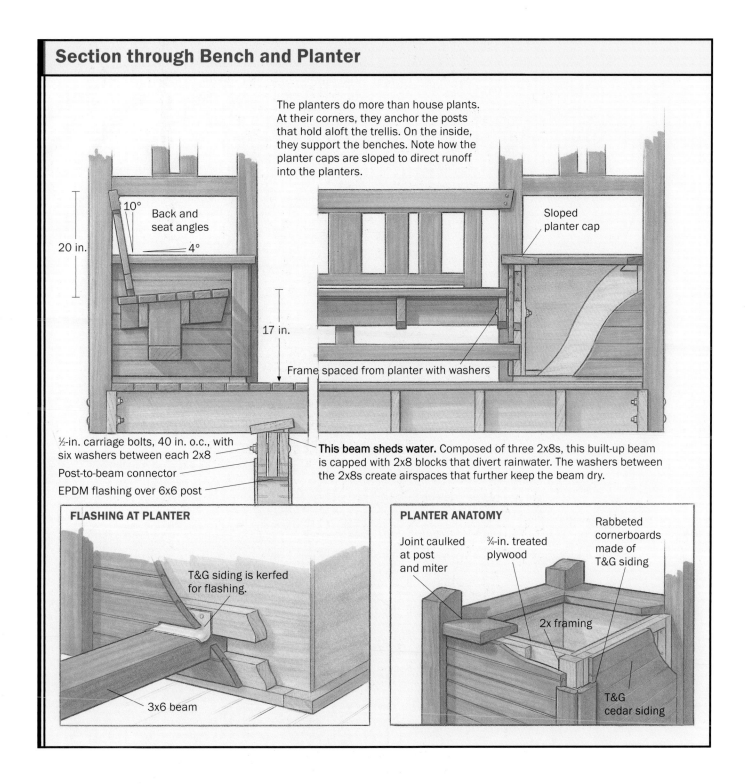

The planters do more than house plants. At their corners, they anchor the posts that hold aloft the trellis. On the inside, they support the benches. Note how the planter caps are sloped to direct runoff into the planters.

10°
Back and seat angles
4°
20 in.
17 in.

Sloped planter cap

Frame spaced from planter with washers

½-in. carriage bolts, 40 in. o.c., with six washers between each 2x8

Post-to-beam connector

EPDM flashing over 6x6 post

This beam sheds water. Composed of three 2x8s, this built-up beam is capped with 2x8 blocks that divert rainwater. The washers between the 2x8s create airspaces that further keep the beam dry.

FLASHING AT PLANTER

T&G siding is kerfed for flashing.

3x6 beam

PLANTER ANATOMY

Joint caulked at post and miter

¾-in. treated plywood

Rabbeted cornerboards made of T&G siding

2x framing

T&G cedar siding

to circulate as well as to make the outer faces of the beam flush with its 6x6 posts. Making the beam the same width as the post allowed us to attach the beam and the post easily with steel connectors instead of toenails.

Before setting the beam on them, however, I covered the top of each post with a piece of EPDM roofing membrane. I also capped the beam with 2x8 blocks between the joists. The blocks are sloped to divert water and dirt off the top of the beam. The blocks also provide secure nailing for the joists. In the spots where I needed doubled joists, I used pressure-treated plywood spacers between them to promote drainage and ventilation. I think precautions such as these are cheap insurance. The fewer paths there are for moisture to get into the framing, the better.

The Planters Are Structural

At first glance, the cross section of the planter boxes might look like a 2x wall sheathed on the inside with treated plywood and on the outside with T&G siding (bottom right drawing, p. 45). But they are plywood boxes covered with framing and siding. The plywood contributes the rigidity; the framing provides a flange for securing the 4x4 posts affixed at the three outer corners of each planter to support the trellis.

Bill and I assembled the boxes with Fastap® self-drilling screws. These screws are coated with a nongalvanized proprietary finish called Duracoat® that is supposed to be more durable than galvanizing. More importantly, the coating isn't corroded either by the copper in pressure-treated wood or by the tanins in red cedar.

To overcome the problem of trying to get the tops of four posts to make a perfectly

straight line where they meet the beam, we installed the two outer posts, then the beam, then the two intermediate posts.

Like the posts, the beams are clear, pressure-treated 4x4 cedar. We anchored the beams to the posts with 8-in. galvanized helix nails (you predrill for these babies). Where the beams intersect each other, the upper beam is notched out to a depth of 1 in. Counterbored screws secure the connection, and the counterbores are filled with teak plugs installed with clear silicone instead of glue.

Next, Build the Benches

The benches are suspended between the planters atop an egg-crate frame of interlocking 2x4s, bolted through the planter walls at the ends and supported in the middle by a 3x6 post (photo, below). The post is screwed from underneath to a 3x6 beam, which also bears on the planter framing. To accommodate the flashing (bottom left drawing, p. 45), this beam had to be installed before the siding went on the planters. But the egg-crate frame was bolted on after the

Keep the water out of the framing. An interlocking frame of 2x4s supports the built-in benches. At midspan, a short 3x6 post supports the frame. The top of the post is capped with metal flashing to shed water, and the bolted supports affixed to the planters are shimmed out with washers to create an airspace.

siding was in place. Once again, I used a half-dozen extra washers to maintain airspace between the frame and the siding.

The 2x4 seat slats for the benches were installed with hidden fasteners called Dec-Klips instead of nails. Dec-Klips are a good way to avoid nailing through the face of the decking, and one of these days I'll use them for a whole deck.

The backs of the benches make a good show of being joinery but actually are put together entirely with screws. I considered using biscuit joinery and waterproof glue, but I shied away from that for two reasons. First, it would have taken longer to do and would have required kiln-dried stock; and second, I didn't trust rigid glue joints to hold up under the kind of flex and stress I knew the back would take.

I made up the backs as ladders around which I assembled the frames. The back slats are 5/4 KD red cedar. All other parts are also clear red cedar but milled from green 2x4 stock selected for dryness as much as for grain pattern and straightness. Toward the end of summer, you can sometimes find some pretty dry wood in the piles out in the yard, especially shorts, which don't sell as fast. I used Fastap screws for these assemblies, drilled and counterbored, and plugged where exposed with teak plugs that I left proud and sanded lightly. (Unless you're going to seal plugs in with paint or varnish, you might as well leave them proud deliberately because if you cut them flush, they'll pop out when they expand.)

The 12-in. Delta® portable planer I bought not long ago got almost as much use as the radial-arm saw on this job. Thicknesses graduate throughout the seat backs, from the full 2x top rail to the $^{15}/_{16}$-in. finished thickness of the 5/4 slats. Shadowlines and stepped joints are a practical and beautiful Greene & Greene legacy for which I thank them daily. The top rail, for example, started out as a 2x4. I planed it smooth, then ripped it to just less than 3-in. The offcut, about ⅜-in. thick, was planed to 1⅛-in. wide and

trimmed to the length between the 1¼-in. thick main uprights. It became the means of fastening slats. The completed seat backs are joined to the benches with screws from underneath and to trellis posts through notched wings at either end of the top rail.

The Planter Caps Are Sloped to Drain

The planter caps are 2x6s tapered in section (bottom right drawing, p. 45). I milled the taper by running the 2x6s through the planer while atop a sloped board. Before planing the slope, I ran drip kerfs in the bottoms of the cap material ½ in. from each edge.

The caps were mitered using biscuit joinery for alignment, and they were fitted tightly around the trellis posts. Then I used a straightedge and razor knife to cut a channel about ³⁄₁₆-in. square through the joint, which I filled with a marine-grade polysulfide caulk, Sikaflex® 231. This caulk comes in the usual three or four colors. I used black—not subtle, but handsome to my mind, and reminiscent of boat decks. This is a time-consuming and finicky detail, but so far (I've used it on only one other deck, built in the spring of 1991) it appears to be successful in eliminating Curling Miter Syndrome, which is so painful and familiar to outdoor woodworkers.

Planters, benches, and the simple lattice between the posts were finished with two coats of clear Ducksback® Total Wood Finish, an exterior finish that goes on milky and dries perfectly clear without appreciably darkening the cedar. The trellis and decking are pressure-treated and were left unfinished. Although you can use Ducksback on treated lumber, too, it's a good idea to let the treatment be absorbed and to let the wood dry for at least two months before application.

All the Doors Fold Outward

Early mornings and late nights during the building of the deck had been given to head-scratching about those bifolding French doors. I had to draw something for our doormaker, Peter Fenger, so he'd know the job was for real and plug it into his schedule. But I changed just about every detail before he built the doors. By that time, when Pete saw me coming through the door, he would get that look on his face. "This is it, Tony," he finally said one morning. "We're cutting."

Here's what we came up with, and why (photo, below). First, the doors had to open onto the deck; the dining room was too small to lose floor area to them. The fold would bring the inside faces of each pair of doors together. To allow them to fold flat without the handle getting in the way, the two center panels are wider than the outside

ones. Serendipitously, this also makes a nicer rhythm of proportion than doors of equal width, I think.

Four doors meant eight stiles taking up glass area, so we didn't want the stiles to be any wider than necessary. At the same time, a sealed unit made with laminated glass on both sides is not light. We settled for 3-in. wide stiles, with a 4½-in. top rail and a 7-in. bottom rail. To add strength and to make room for a ½-in. airspace between the panes, we made the sash 2¼-in. thick. The additional thickness also allowed us to use 4x4 hinges and still have plenty of wood for a rabbeted astragal at each meeting stile.

The biggest problem was closure hardware. How to pull four doors tight against their jambs without having handles at the folding hinge point? Large surface bolts were an option, but I thought it would be cumbersome to have to lock each door independently, top and bottom. Besides, even though the entry door in the adjacent wall

Inset straps align the doors. Cocobolo straps fit into grooves in the door frames. When the doors are folded, the straps are retracted. Note how the doors are slightly different widths, which allows the doors to fold flat to one another without the doorknob getting in the way.

meant the hardware didn't have to be designed for constant use, I wanted it to be clear what a person had to do to open the doors. A profusion of bolts seemed likely to confuse. Dummy handles on the inactive leaf of any pair of French doors contradict this principle, too, by offering an option that turns out to be false. How many times do we need to suffer that small embarrassment? If there is only one handle to grasp, on the other hand, there can be no confusion.

One handle, large enough to grasp comfortably, could be provided only by a cremone bolt because the door stile isn't wide enough for any standard lockset. The principle of uncluttered surfaces and transparency of function suggested ordinary flush bolts in the edge of the inactive leaf. This combination meant the need of finding a way to lock each pair of doors into a single panel so that each could be opened and closed as a single door. I didn't know of any commercially available hardware capable of doing this, so I designed some wooden strap bolts for the job (photo, facing page). They fit into grooves cut into the rails and stiles at the tops and bottoms of the doors.

Time to Cut the Grooves

Plowing four sizable grooves in a set of $1,800 custom-made doors seemed a pretty earnest commitment to an untried design when it came time to do it, and I will admit to a degree of nervous procrastination over the job. The night before, I made Bill look at mock-ups screwed to the doors where the bolts were to go to help me determine the exact length and width. As always, he was patient, optimistic, and gently biased

toward optimum mechanical efficiency. We settled on a width of 1⅛ in. and a length of 13¼ in., the maximum we could get without quite intersecting the line of the primary hinge stile.

I also wanted to make sure the doors were fitted to their openings before cutting the grooves for the bolts, so I installed spring bronze weatherstripping from Pemko Industries at the head and side jambs and between each door. An aluminum and vinyl door bottom (also from Pemko Industries) was fitted in the groove I had asked Peter to machine. Before installing the weatherstripping, I sealed with oil all areas that couldn't be reached after the weatherstrip was in place.

I made the bolts and their retaining straps from cocobolo, a dense, hard wood that grows in Mexico. Each pair of doors was removed and laid on sawhorses to rout the slots, but the final fitting of bolts had to be done with doors in place. Achieving the right degree of resistance meant balancing the pressure of the weatherstripping by easing the inside surfaces of the retaining straps with sandpaper. The beauty of surface mounting retaining straps with countersunk brass screws is that if the bolts cause problems, it will be simple to make adjustments.

Tony Simmonds operates DOMUS, a design/build firm in Vancouver, British Columbia.

Sources

Ben Manufacturing Inc.
P. O. Box 51107
Seattle, WA 98115
206-776-5340
Dec-Klips

Fastap
13909 NW Third Court
Vancouver, WA 98685
800-874-4714
www.fastapscrews.com

Masterchem Industries
P. O. Box 368
Barnhart, MO 63012
800-325-3552
www.masterchem.com
Ducksback Total Wood Finish

Pemko Industries
P. O. Box 3780
Ventura, CA 93006
800-283-9988
www.pemko.com
Spring bronze weatherstripping

Sika Corp.
201 Polito Ave.
Lyndhurst, NJ 07071
800-967-7452
www.sikaonstruction.com
Sikaflex 231

Adding a Sunroom with Porch

■ BY DIDIER AYEL

I've seen too many old homes ruined by inappropriate additions. In some of those cases, it would have been better to tear down the house rather than graft an unsuitable addition to it. I'm not saying an addition has to be a slavish copy of the original. But it must fit. It must add to the original, not just be added to it.

So when I was asked to design a sunroom and porch addition for a classic Victorian in Montreal, the challenge was to make the addition appropriate to the house and useful to the clients. By the time I met Dennis and Suzanne, I was already familiar with their house, which is one of the few examples of late Victorian in the neighborhood. On the back of their house, which faces west and the couple's lovely garden, a small two-story shed addition had reached the end of its life span (inset photo, facing page). The shed was a light frame structure with single-pane windows and, of course, no insulation. The enclosed lower portion of the addition was used for storage. The upstairs balcony was accessible from the second floor.

Dennis and Suzanne wanted to replace the shed with a Victorian-style sunroom covered with a second-story porch. They wanted a powder room in the addition, a full-height basement beneath it, and outside access to the basement, which already was accessible from the main basement. They also wanted the addition to be as maintenance-free as possible.

The Design Had to Suit the House

I decided the sunroom should be designed like a porch, supported by classical columns yet enclosed with large windows and molded panels on three sides. In the center, French doors would open onto a stair leading to the garden.

To prevent obstructions to the view from the addition, we would put the powder room in the corner of the sunroom against the house. To respect the exterior design and the continuity of windows, columns and molded panels, the powder-room partition

A fitting replacement for a tired addition. An outdoor porch for warm weather opens to the upstairs hall of the main house, while below it, the sunroom is a comfortable setting for all weather. The original addition, right, was flimsy, old, and past its prime.

Lots of light for all seasons. When cold weather settles into a long, dark winter, this sunroom is cozy and full of light.

would intersect the exterior wall between two windows so that it remained invisible from outside.

To help relate the addition to the house, the overhang above the sunroom would be supported by wide brackets, similar to the brackets beneath the cornice returns on the main house. Victorian brackets and moldings also would enhance the design for windows, transoms, and second-floor columns.

Finally, transoms above the windows and French doors—each topped by a molded panel—would create the illusion of height necessary to match the existing house and to keep all openings in proportion.

The most significant change over the old shed was the enlargement of the basic footprint. The previous 20-ft. by 12-ft. sunroom with four windows and a patio door on the facade became a 27-ft. by 14-ft. structure with six windows across the facade—three on each side of the French doors—and four windows on each side (photo, p. 51). Also,

instead of a stair running perpendicular to the sunroom as in the old addition, we designed a small landing in front of the French doors with steps coming off each side leading to the garden. The result was only slightly more expensive but much more elegant.

Ensuring a Watertight Foundation

The new basement wall would be 10 in. thick and have an average height of 5 ft. above the finished grade. Because we needed 7 ft. of headroom in the basement, the concrete slab was placed 2 ft. below grade. To protect the footings from frost, they sit 2 ft. 6 in. below the level of the slab, or 4 ft. 6 in. below grade, which is the minimum depth in Montreal.

To provide a tight, dry connection between the new and old foundation walls, we cut a 4-in.-deep vertical groove in the old stone wall, into which we installed a PVC

Waterproofing the Joint between New and Old Foundations

A PVC spline spans the joint between new foundation and old. To install the spline, two vertical 4-in.-deep kerfs were cut in the old foundation wall and the thin section between the kerfs removed. The PVC spline was grouted into the enlarged kerf using epoxy concrete. To help ensure a watertight joint, 1x1 keys were nailed to the inside of the forms where they met the old wall. When the forms and the keys were removed, the keyways were filled with epoxy concrete.

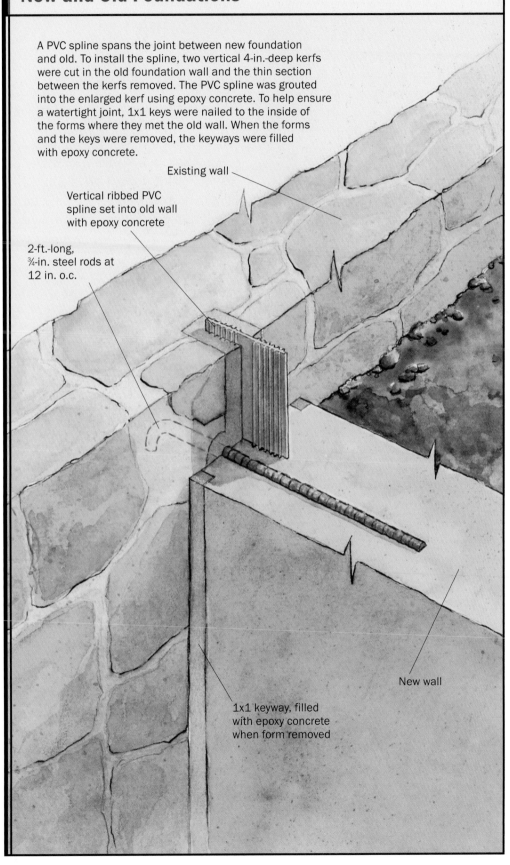

Existing wall

Vertical ribbed PVC spline set into old wall with epoxy concrete

2-ft.-long, ¾-in. steel rods at 12 in. o.c.

New wall

1x1 keyway, filled with epoxy concrete when form removed

spline (drawing, p. 53). The ribbed spline, called a DuraJoint® waterstop, was grouted in place with epoxy concrete, which was allowed to harden before we poured the new foundation over it. With the 9-in.-wide waterstop embedded halfway into both the old and new foundation walls, I thought the joint would be watertight.

We also installed hooked steel rods between the two walls—2-ft. lengths of ¾-in. rebar at 12 in. o.c.—which ran beside the flexible spline the height of the wall. We nailed 1x1 strips to the inside of both interior and exterior wall forms. After removing the forms, we filled the 1x1 spaces with epoxy

concrete. This measure would provide even greater insurance against water intrusion.

Protecting the Sunroom from Heat and Moisture

The summer porch on the second floor was open to the weather, so we had to waterproof its deck carefully to protect the sunroom below. The deck system consists of ⅝-in. pressure-treated plywood sheathing nailed to the sunroom's ceiling joists, then covered with the built-up asphalt roofing.

Solitude on the porch. Four skylights open the porch to sunlight; a deck over a waterproof membrane protects the sunroom below.

Perpendicular to the house, wide strips of foam mudsill gasket material run over the roofing every 16 in. o.c.; over that, we set 2x4 sleepers, which are held in place by the weight of the decking. The foam would protect the membrane. We cut slots across the underside of each sleeper at 16 in. o.c. to help the water move in every direction. Finally, the 2x6 treated decking was nailed across the sleepers and spaced about ⅜ in. apart.

To secure a watertight joint between the deck and the house, we raked out the mortar joints between the bricks, then glued and nailed a piece of ½-in. plywood to the house. After the asphalt membrane was installed up and over the plywood, we embedded horizontal aluminum flashing in the mortar joints to protect the edge of the membrane, which rides 6 in. up the masonry wall.

A Hip Roof with Skylights

The choice of a hip roof was evident from the beginning of the preliminary design. An attic window in the gable of the house overlooking the backyard prevented us from building the addition too high. Besides, given the size of the addition, a gable roof would have appeared too tall for the house.

Before construction on the addition began, the clients decided to reroof their house with blue metal standing-seam roofing, which we used to roof the addition. The new roofing makes a strong visual tie between the new and the old.

Originally, I planned to have a flat ceiling over the sun porch made of ¾-in. tongue-and-groove pine beadboard. However, we decided to take advantage of the hip-roof slopes and to fasten the porch ceiling directly to the rafters, adding four skylights, two in the center hip and one on each side. That change provides a much more interesting look inside the porch. Each skylight measures 2 ft. by 6 ft. and provides plenty of additional light on the deck (photo, facing page).

Man-Made Details Save Time and Money

The exterior finish was important to the project. For the large exposed foundation walls, natural-stone veneer was too expensive, and neither stucco nor brick was satisfying. The solution was cultured stone. This artificial stone is much lighter and cheaper than natural stone. Natural stone would have been about $20 per sq. ft. The cultured stone cost about $12 per sq. ft.

All molded panels and brackets were made by Fypon® Molded Millwork®. This company offers a wide selection of molded millwork made from high-density polymer that can be used like wood millwork, but it comes at a much lower price.

Dennis and Suzanne wanted the addition to be as maintenance-free as possible, so they asked for a balustrade that was made of PVC. I wasn't crazy about the idea, but the difference between wood and plastic is almost invisible. I usually prefer painted wood to PVC, but this balustrade really looks like a wooden one. For architectural unity, the handrail and the balusters along the stairs leading to the garden were made of the same material.

When Dennis and Suzanne replaced existing windows of the house, they chose vinyl windows from M.Q. Windows. Made in Canada with a German technology, the windows—unlike most vinyl products—have the beautiful proportions of wood windows, but they are almost maintenance-free. We decided to use the same type of doors and windows for the sunroom. We chose casement windows with double-insulated glass and real dividers. The French doors also have real divided lites.

Didier Ayel is an architect in Montreal, Canada.

Sources

Cultured Stone®
P. O. Box 270
Napa, CA 94559
800-255-1727
culturedstone.com

Fypon Molded Millwork
P.O. Box 238
Seven Valleys, PA 17360
800-955-5748
fypon.com

M.Q. Windows
1855 Griffin Road, #A-274
Dania, FL 33004
954-929-8500

A Classic Sunroom

■ BY STEVEN GERBER

There's an awkwardness about the so-called sunroom additions typically stuck onto houses these days. These additions are usually factory-made greenhouses either with hipped or curved glazing that gives a house that certain fast-food-restaurant look. Most sunrooms don't relate to the houses they're attached to and have little or no identity. So when my clients approached me with the idea of transforming an existing screened porch into a year-round sunroom, I wanted to design a space that looked as if it had always been part of the existing house.

Columns Allow for Large Windows

The house sits on a small knoll that is approached from below with the house angled toward the street to provide access to a gable-end basement garage. The sunroom was to be built off the corner of the house just above the garage, making the sunroom a highly visible part of the house's most prominent facade. It was imperative that the sunroom's design complement the architectural qualities of the house.

I modeled my design on the sunrooms and greenhouses found in many grand period houses with classical origins built in this country before World War II. The open spaces between the columns supporting the roofs of these structures naturally lend themselves to considerable window area. The roof glazing could be limited without losing the open feeling. My challenge was to design an aesthetically pleasing room with large areas of glass, but one that would be comfortable and energy-efficient year-round. A deck built into the corner formed by the house and the sunroom would extend the sunroom spaces to the outdoors during the warmest months.

Once I'd resolved to use the column or pilaster as an organizing element, I quickly sorted through my basic options as to how many columns there would be on each side, whether they would be equally spaced, and so on. In the interest of budget, we decided early on to use standard-size windows and to design around those dimensions. We chose Pella® windows, which had the proportions I wanted as well as a low-maintenance factory-painted aluminum exterior cladding. The aluminum-clad windows also have a thin exterior casing that worked well with my custom exterior trim.

Instead of the large expanses of roof glazing that most sunrooms sport, I opted for a well-insulated roof with a few large skylights. These skylights allow for a visual connection to the trees and sky outside, and limiting the overhead glazing helps to keep

Classic exterior. Built-up panels and moldings create a formal facade that complements the architectural qualities of the house.

Sources

Green Mountain Precision Frames
P. O. Box 293
Windsor, VT 05089
802-674-6145

Pella Corporation
102 Main St.
Pella, IA 50219
800-547-4758
pella.com

Plastmo Ltd.
255 Summer Lea Rd.
Brampton, Ontario
CANADA L6T 4T8
888-793-9462
plastmo.com

the room from frying in the summer and freezing in the winter. The skylights have electronically controlled blinds and motorized openers that also help to control the room's climate.

Invisible Timber Connectors Hold the Trusses Together

I had a couple of areas of concern regarding the structure of the sunroom. First, because of the large amount of wall area taken up by windows, I worried about the horizontal shear strength of the walls. To solve this problem, I made the plywood sheathing continuous from the area below each window to the small sections between the windows. The natural tendency in this case would have been to cover these areas with separates strips of plywood. To achieve the required shear strength, the nails had to be spaced more closely together at the edges of the plywood, especially on the gable-end wall, which doesn't benefit from being attached directly to the house for any sort of lateral strength.

Another even more challenging problem was creating a roof structure for the cathedral ceiling. Because the roof framing would be fully exposed inside the sunroom, I decided to support the roof with timber trusses. The problem was how to connect the members.

I looked at several connection options, including mortise-and-tenon joints, metal plates with through bolts, and metal connectors. Even though the trusses were made of 6x6 Douglas fir, I worried that a mortise-and-tenon joint wouldn't leave enough meat on the end of the horizontal member to resist the outward thrust of the roof. Plates and through bolts are clunky and wouldn't be in character with the sunroom's interior, so I decided to use timber connectors to join the trusses.

I ended up having the roof trusses made by Green Mountain Precision Frames. They prefabricated the four roof trusses using their Timberlok Joinery System, a concealed, practically invisible steel-connection system.

An Informal Interior

I wanted the sunroom's interior to have a less formal feel than its classical exterior, so I began at the top with a natural-wood ceiling (photo, facing page). If we had sheathed the roof with standard plywood, the spacing of the roof trusses would have required additional purlins. So I used 2x6 tongue-and-groove Douglas fir instead. The 2x6 boards easily spanned the distance between the trusses, and they added beautiful color and texture to the ceiling at the same time.

To complete the roof, the Douglas fir was covered with an air barrier, a 4-in. layer of foil-faced foam insulation, 2x2 furring strips (to create a ventilation cavity) and plywood sheathing. The plywood was topped with self-adhesive rubberized membrane and finally a layer of asphalt shingles. I decided to forego the usual ridge vent for aesthetic reasons. Continuous soffit vents, a low slope (3-in-12), and good crosswinds have kept the roof dry through the extremes of the past two New England winters.

As a counterpoint to the natural wood above, I decided to go with a stone floor. I chose an Indian slate for its deep silvery-blue color and natural cleft finish. In addition, the slate feels cool in the summer and stores heat during sunny winter days. The ½-in.-thick pieces of slate were set into a ½-in. mud bed on top of the ¾-in. plywood subfloor. The mud bed let me use larger pieces of stone and added to the thermal mass of the floor.

Exterior Panels Made in Layers

The most enjoyable part of the project was detailing the exterior trim. For cost reasons I opted for moldings that could be made on site. I specified redwood for most of the

Exposed trusses and a wood ceiling give the interior a more relaxed feeling than the exterior's more formal features.

exterior trim because redwood resists decay and holds a paint finish.

Mark Iverson of Jolin Construction in Dedham, MA, made the flat panels below each of the windows out of medium-density overlay board. He applied layers of trim on top of the board to create the recessed panel (bottom left photo, p. 60). The pilasters between the windows were given a similar treatment. All critical joints in the trim were biscuit-joined to ensure that they would not open. All the trim on the outside corners of the sunroom was mitered and glued to provide a seamless, solid appearance, and I made sure that all the trim was back-primed before it was installed.

I had hoped to put copper gutters and downspouts on the sunroom, but I found that the cost was exorbitant. The standard aluminum gutters that I looked at were out of proportion with the room. So instead I chose a less expensive PVC-gutter system made by Plastmo that was beautifully designed and just the right size (top left photo, p. 60).

The small round window on the pediment was the only custom window in the sunroom. The 7-in. radius was not a size made by any of the major manufacturers, so I had it fabricated at a local shop. The round window adds a finishing touch to the facade and creates a special light feature in the interior as well.

Details of Classic Attire

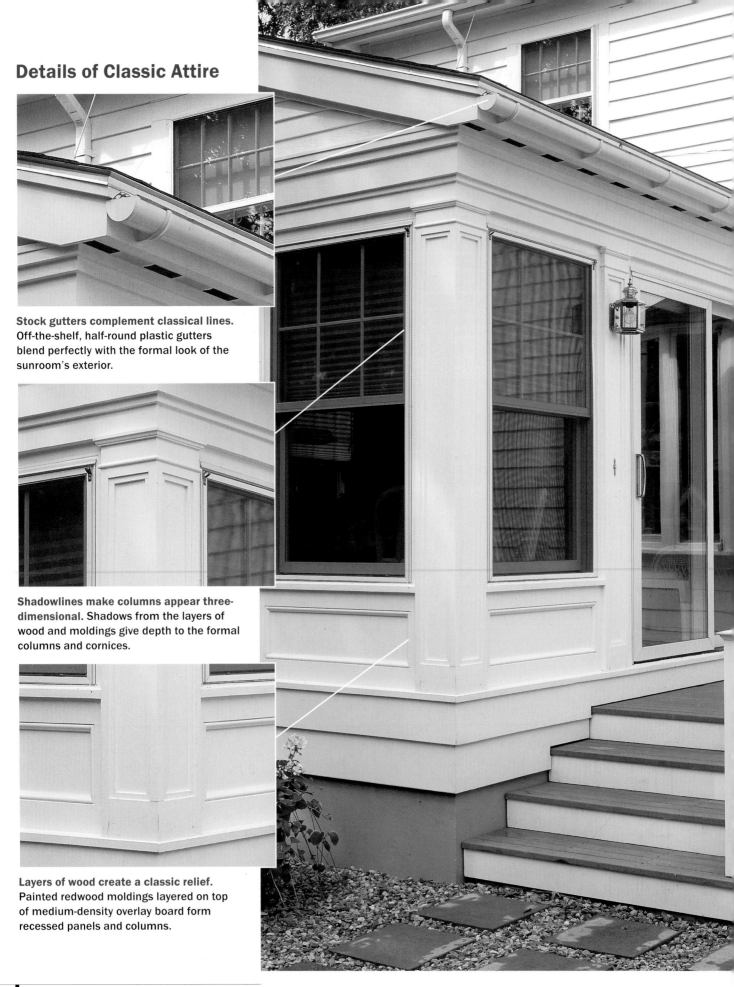

Stock gutters complement classical lines. Off-the-shelf, half-round plastic gutters blend perfectly with the formal look of the sunroom's exterior.

Shadowlines make columns appear three-dimensional. Shadows from the layers of wood and moldings give depth to the formal columns and cornices.

Layers of wood create a classic relief. Painted redwood moldings layered on top of medium-density overlay board form recessed panels and columns.

Creative Detailing Means a Longer-Lasting Deck

I designed the adjoining deck to complement the sunroom and to extend its spaces outdoors. The deck is supported by 4x4 pressure-treated posts topped with a decorative cap and wrapped in painted redwood above the deck boards. The railing pickets are all mortised into the top and bottom rails, which are attached to the posts with concealed metal angles and screws.

A redwood lattice skirt wraps around the deck to create a more finished look and to keep leaves and debris from blowing underneath. One of the lattice panels is hinged for access to the crawlspace under the deck. Another crawlspace under the sunroom is also accessible through a removable louvered panel in the foundation wall of the sunroom.

Typically, when a deck is built, sheathing or siding on the adjoining side of the house is often trapped behind the framing of the deck. This wood can't be repainted or inspected for rot without removing part of the deck. I stepped the sunroom foundation so that there is only concrete behind the deck frame (drawing, below). The sunroom's floor joists rest on a shelf that is formed on the inside of the stepped wall. This detail puts all of the painted surfaces above the deck.

The deck framing doesn't actually touch the sunroom foundation wall. But along the existing wall of the house, where the deck joists are supported, I removed the painted clapboards that would have been concealed behind the rim joist and had aluminum flashing wrapped over the top of the joist. I also left a small ⅛-in. gap between the stained surfaces and the painted surfaces. The gap makes it easier to paint and repaint without leaving a sloppy edge.

Steven Gerber is an architect based in Brookline, Massachusetts.

Foundation Is Built Up for a Better Deck Detail

A section of the sunroom foundation beside the deck is raised so that no wood on the sunroom walls is left behind the deck framing to deteriorate. Sunroom-floor framing lands on a built-in shelf.

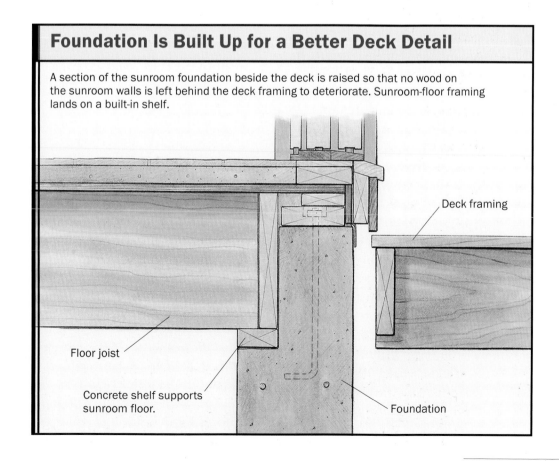

Deck framing

Floor joist

Concrete shelf supports sunroom floor.

Foundation

Keeping a Dormer Addition Clean and Dry

■ BY NICHOLAS PITZ

Retrofitting a shed dormer in an occupied house can be a disruptive project. But with careful planning, intrusions such as a constant parade of workers and demolition debris can be kept out of the house. And although there's always some time when the roof is open or a wall is missing, avoiding weather damage is straightforward: Just expect pouring rain every night, and plan accordingly.

First, Build the New Roof

The work on this house involved replacing two small existing dormers with a large shed dormer, and the job affected almost every upstairs room. To ensure that the project was weathertight, we'd build the new dormer's roof first, but we needed room for demolition and reconstruction under it.

We began by framing the dormer's outside wall with four posts and a long header. This approach meant we could frame the dormer roof while cutting only four small holes in the original roof. Then, after we built the new roof and removed the old dormers, we could fill in the studs and the window framing.

We marked the locations of the wall posts to correspond with where we wanted the interior partitions to fall. Then we cut out small sections of roof with a reciprocating saw to give us access to the top plate of the existing wall and installed the first corner post. When this post was carefully braced and plumbed, we used a water level to calculate the height of the other corner post to ensure that the roof was level. Once this second post was positioned and plumbed, we strung a line between the two end posts and set the remaining posts to that height, making sure the load was transferred directly to a stud below.

The small dormers had to go.
The original house lacked
space upstairs, so a big shed
dormer was needed. To minimize disruption for the folks
living in the house, access to
the work zone was via ladders
and scaffolds, not through
the house.

With the posts braced and in place, we installed the headers. At the end of the day, we covered the holes in the roof made by the new posts with aluminum flashing and plastic sheeting (left drawing, p.65).

Tear Out Only What Can Be Rebuilt That Day

We framed the new dormer's roof in two sections over two days. We calculated where the new dormer rafters would intersect with the oldest section of roof, cut a 2-ft.-wide section, and removed it. This approach gave us access to the attic without having to go through the house, and also provided light and ventilation to the attic where we would be working.

After the new rafters were securely in place, the sheathing and the #30 felt were installed. The next day, we repeated the performance for the rest of the dormer roof. We now had a reasonably weathertight lid on the work area and could begin the demolition process. To keep out rain, we protected the work site with tarps, which were layered like roofing shingles and held fast with furring strips (right drawing, p. 65).

Staying Ahead of the Weather

To keep out rain, build the new roof before removing the old one. Supporting the new dormer roof with one long header and four posts meant cutting only four small holes in the old roof. (1) Setting and temporarily flashing the posts was the first day's work. Half of the dormer's new roof was framed and then sheathed the next day, (2) and all the openings were flashed and tarped against the wind and rain. The dormer's roof was completed and sheathed, and #30 felt paper kept out the rain. (3)

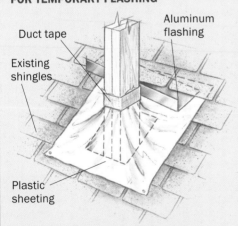

DUCT TAPE AND PLASTIC FOR TEMPORARY FLASHING

Duct tape

Aluminum flashing

Existing shingles

Plastic sheeting

Flashing is slipped under the shingles, bent, and canted at an angle to divert water around the post. Heavy plastic sheeting (6-mil poly) is slid under the flashing, wrapped around the post, and tacked down with roofing nails. Duct tape seals the poly to the posts.

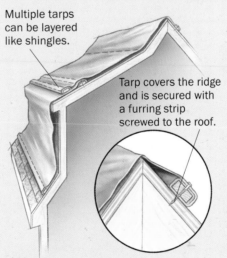

HAVE TARPS IN PLACE BEFORE IT RAINS

Multiple tarps can be layered like shingles.

Tarp covers the ridge and is secured with a furring strip screwed to the roof.

If the tarp is in the way of the work, then secure the top edge to the roof, roll up the tarp around a 1x3 furring strip, and tie the rolled-up tarp in place. When the rain starts, cut the ties.

Keeping a Project Neat

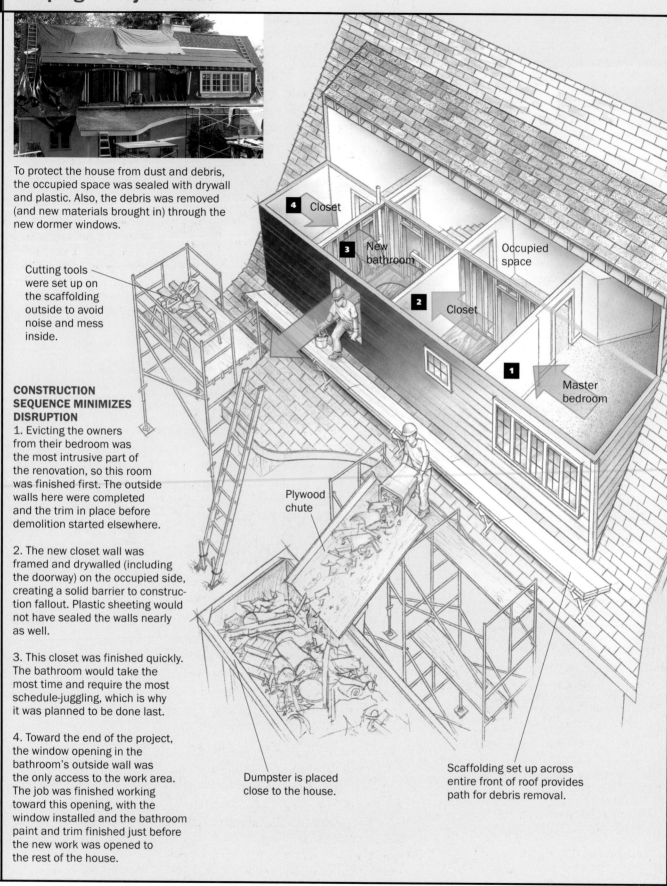

To protect the house from dust and debris, the occupied space was sealed with drywall and plastic. Also, the debris was removed (and new materials brought in) through the new dormer windows.

Cutting tools were set up on the scaffolding outside to avoid noise and mess inside.

4 Closet

3 New bathroom

Occupied space

2 Closet

1 Master bedroom

Plywood chute

Dumpster is placed close to the house.

Scaffolding set up across entire front of roof provides path for debris removal.

CONSTRUCTION SEQUENCE MINIMIZES DISRUPTION

1. Evicting the owners from their bedroom was the most intrusive part of the renovation, so this room was finished first. The outside walls here were completed and the trim in place before demolition started elsewhere.

2. The new closet wall was framed and drywalled (including the doorway) on the occupied side, creating a solid barrier to construction fallout. Plastic sheeting would not have sealed the walls nearly as well.

3. This closet was finished quickly. The bathroom would take the most time and require the most schedule-juggling, which is why it was planned to be done last.

4. Toward the end of the project, the window opening in the bathroom's outside wall was the only access to the work area. The job was finished working toward this opening, with the window installed and the bathroom paint and trim finished just before the new work was opened to the rest of the house.

Keep the Work Site Separate from the House

The interior of the dormer was divided into four areas: master bedroom, closet, the daughter's bathroom, and her closet. It was important to complete the master bedroom quickly, so we built the front wall, sheathed it, installed the window, and had it drywalled and trimmed before some of the other sections were even started. By finishing this room first, the owners got their bedroom back quickly.

We were careful to isolate the workspace from the rest of the house. Two small existing closets were turned into walk-in closets (drawing, facing page). Before we commenced with serious demolition of either closet, we reframed the door openings, drywalled over them, and taped the seams, planning to cut out the doorways after the dirty work was done. Having the living side of the wall closed acts as a barrier against construction while allowing the electricians and plumbers to do their work on the other side.

Even the small amount of drywall we did on the living side of the wall meant dust. We took the usual precautions. A plastic runner protected the upstairs carpet. To contain airborne dust, I tacked tarps across doorways, placed a box fan in a window as an exhaust, and vacuumed at the end of each workday. On the other side, we completed almost all the work before we cut through to install the new doors and trim.

Because the bathroom door wasn't being moved, we taped it shut and put a sheet of foam insulation over it for protection. The bathroom then was gutted and rebuilt.

No Muss, No Fuss

Much of this job revolved around not making a mess. We planned the placement of the dumpster so that we could build a plywood chute to it for debris removal, but

Tips for Any Remodel

- Minimize traffic through the house.
- Scrupulously seal off the living space from the work zone.
- Use a window fan in the work zone to exhaust dust and fumes.
- Take extra precautions to protect the stuff that stays. If the floor stays, for instance, cover it with plywood (and tape the seams) rather than rely on a tarp.
- Clean up the work zone at the end of every day.
- To keep peace in the house, consider hiring a cleaning service halfway through the project.

Completed with careful attention to cleanliness and the homeowner's privacy, this dormer addition ends with appreciative customers and an enhanced reputation for the builder.

kept it out of the way of deliveries. We were working directly over the driveway, and we made sure we swept up carefully every night.

The bathroom's outside wall was left unframed as an access to the second floor, and consequently, little traffic went through the house. The dormer was finished toward that opening, and the day before the tile was installed in the bathroom, we finished framing the rough opening and hanging the drywall, then installed the window and trim. The tilesetter's wet saw was set up on the scaffold outside, and his helper spent the day getting a tan and passing cut tiles through the window.

Every remodeling project is an intrusive ordeal for the homeowner. The single most important aspect of our remodeling strategy was that my clients appreciated the efforts we took to protect their home and privacy.

Nicholas Pitz is the principal of Catamount Design and Construction in suburban Philadelphia.

A Different Approach to Rafter Layout

■ BY JOHN CARROLL

Earlier this year, I helped my friend Steve build a 12-ft. by 16-ft. addition to his house. A few days before we got to the roof frame, I arrived at his place with a rafter jig that I'd made on a previous job. I'm a real believer in the efficiency of this jig, so I told Steve that it would enable me to lay out the rafters for his addition in 10 minutes.

His look suggested that I had already fallen off one too many roofs. "Come on, John," he said. "Ten minutes?" I bet him a six pack of imported beer, winner's choice, that I could do it.

Like most builders, I have a long and painful history of underestimating the time different jobs require. In this case, however, I was so certain that I agreed to all of Steve's conditions. In the allotted time, I would measure the span of the addition; calculate the exact height that the ridge should be set; measure and mark the plumb cut and the bird's mouth on the first rafter; and lay out the tail of the rafter to shape the eaves.

When the moment of reckoning arrived, we set a watch, and I went to work. Eight minutes later, I was done and in the process secured the easiest six bottles of Bass® ale in my life. With this layout in hand, we framed the roof in 5½ hours.

Why Should It Take an Hour to Do a Ten-minute Job?

That evening, as we enjoyed my beer, Steve's wife asked him how long he would have taken to do the same layout. "An hour," he said, "at least." Steve is a seasoned carpenter who now earns a living as a designer and construction manager. So why does a ten-minute job require 60, or possibly 90, minutes of his time? The answer is that Steve, like many builders, is confused by the process.

Laying out rafters doesn't need to be complicated or time-consuming. Using a jig makes the process quick and easy.

Rafter jig doubles as a cutting guide. Scaled to the 12-in-12 roof pitch, and with a 1x3 fence nailed to both sides of one edge, this plywood jig is used to lay out plumb and level cuts on the rafters. Notice that the fence is cut short to make room for the circular saw to pass by when the jig is used as a cutting guide (photo, below). The jig will be used later to lay out plywood for the gable-end sheathing.

The first framing crew I worked with simply scaled the elevation of the ridge from the blueprint and then installed the ridge at that height. Once the ridge was set, they held the rafter board so that it ran past both the ridge and the top plate of the wall, then scribed the top and bottom cuts. Then they used this first rafter as a pattern for the rest. This technique worked. And because it's so simple and graphic, I'm convinced that it still is a widespread practice.

There are several reasons why I retired this method decades ago. To begin with, I haven't always had a drawing with an elevation of the roof system, which means that I couldn't always scale the height of the ridge. Second, it's just about impossible to scale the ridge with any degree of precision. Because of this fact, these roofs usually end up merely close to the desired pitch. Third, this method typically leaves the layout and cutting of the rafter tail for later, after the rafters are installed.

My technique is also different from the traditional approach espoused in most carpentry textbooks, which I've always found to be obscure and confusing. In rafter-length manuals and in booklets that come with rafter squares, dimensions are generally given in feet, inches, and fractions of inches. I use inches only and convert to decimals for my calculations.

Another difference in my approach is the measuring line I use. As the drawing on the facing page shows, the measuring line I employ runs along the bottom edge of the rafter. In contrast, most rafter-length manuals use a theoretical measuring line that runs from the top outside corner of the bearing wall to a point above the center of the ridge.

A final thing that I do differently is use a site-built jig instead of a square to lay out the cuts on the rafter (photos, p. 69).

There are lots of ways to lay out rafters, and if you already have a method that works well, your way might be faster than mine. But if you've always been vexed by roof framing, I think you'll find my way easier to understand than most.

Run and Roof Pitch Determine the Measuring Triangle

Framing a roof can be a little intimidating. Not only are you leaving behind the simple and familiar rectangle of the building, but you're starting a job where there is a disconcerting lack of tangible surfaces to measure from and mark on. Most of this job is done in midair. So where do you start?

There are only two things you need to lay out a gable roof. One is a choice of pitch, and the other is a measurement. Usually the choice of pitch was made before the foundation was poured. The measurement is the distance between the bearing walls (drawing, facing page).

After taking this measurement, deducting the thickness of the ridge and dividing the remainder in half, you have the key dimension for laying out the rafters. This dimension could be called the "run" of the rafter; but because it is slightly different from what is called the run in traditional rafter layout, I'll use a different term. I'll call it the base of the measuring triangle. The measuring triangle is a concept that I use to calculate both the correct height of the ridge and the proper distance between the top and bottom cuts of the rafter, which I call the measuring length.

Use the Measuring Triangle to Find the Rafter Length

1. *Find the base of the measuring triangle (the run of the roof).* Measure between bearing walls and subtract width of ridge, then divide by 2. (271-1.75 = 269.25; 269.25÷2 = 134.63).

2. *Find the altitude of the measuring triangle (ridge height).* Divide the base of the measuring triangle by 12 and multiply the result by the rise of the roof pitch. For a 12-in-12 pitch, the base and the altitude are the same. (134.63÷12 = 11.22; 11.22x12 = 134.63). (Editor's note: To present a set of consistent figures, we rounded to 134.63.)

3. *Find the hypotenuse (rafter length).* Divide the base of the triangle by 12 and multiply the result by the hypotenuse of the roof pitch, which is listed as length of common rafter on the rafter square (photo, p. 74). (134.63÷12 = 11.22; 11.22x16.97 = 190.40).

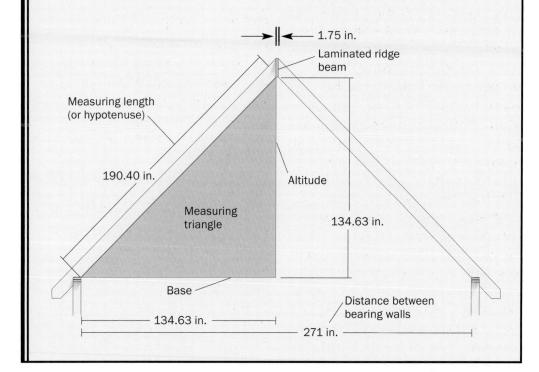

1.75 in.

Laminated ridge beam

Measuring length (or hypotenuse)

190.40 in.

Altitude

Measuring triangle

134.63 in.

Base

Distance between bearing walls

134.63 in.

271 in.

Working with this measuring triangle takes a little getting used to. The biggest problem is that when you start the roof layout, only one-third of the measuring triangle exists. As we've just seen, you find the base of the measuring triangle by measuring existing conditions. Then you create the altitude and the hypotenuse (same as the measuring length) by using that base and some simple arithmetic.

To see how this works, let's look at the steps I followed to frame the roof of a 12-ft. by 24-ft. addition I recently finished.

Build the Rafter Jig First

While the rest of the crew finished nailing down the second-floor decking, I began fabricating a rafter jig based on the pitch of the roof (top photo, p. 69). The desired pitch was 12-in-12 (the roof rises 12 in. vertically for every 12 in. of run horizontally). The basic design of this jig was simple, and the cost was reasonable: three scraps of wood and ten minutes of time.

The first step in making this jig was finding a scrap of plywood about 30 in. wide, preferably with a factory-cut corner. Next, I measured and marked 24 in. out and 24 in. up from the corner to form a triangle. After connecting these marks with a straight line, I made a second line, parallel to and about 2½ in. above the first—the 1x3 fence goes between the lines—then cut the triangular-shaped piece along this second line. To finish the jig, I attached a 1x3 fence on both sides of the plywood between these two lines. I cut the fence short so that it didn't run all the way to the top on one side. That allowed me also to use the plywood as a cutting jig running my circular saw along one edge without the motor hitting the fence and offsetting the cut (bottom photo, p. 69).

Here I should pause and note an important principle. The jig was based on a 12-in-12 pitch, but because I wanted a larger jig, I simply multiplied both the rise and run figures by the same number—two—to get the 24-in. measurement. This way, I enlarged the jig without changing the pitch. This principle holds true for all triangles. Multiply all three sides by the same number to enlarge any triangle without changing its proportions, its shape, or its angles.

To use this jig, I hold the fence against the top edge of the rafter and scribe along the vertical edge of the plywood jig to mark plumb lines and along the horizontal edge to mark level lines.

There are at least four reasons why I go to the trouble to make this jig. First, I find it easier to visualize the cuts with the jig than with any of the manufactured squares made for this purpose. Second, identical layouts for both the top (ridge) and bottom (eave) cuts can be made in rapid succession. Third, I use the plumb edge as a cutting guide for my circular saw. Finally, I use the jig again when I'm framing and sheathing the gable end, finishing the eaves and rake, and installing siding on the gable. I also save the jig for future projects.

Step One: Determining the Base of the Measuring Triangle

The base of the measuring triangle is the key dimension for roof layout. In my system, the base of the triangle extends from the inside edge of the bearing wall to a point directly below the face of the ridge. In this addition, the distance between the bearing walls was 271 in. So to get the base of the measuring triangle, I subtracted the thickness of the ridge from 271 in. and divided the remainder by 2. Because the ridge was 1¾-in. thick laminated beam, the base of the measuring triangle turned out to be 134.63 in. (271-1.75 = 269.25; 269.25÷2 = 134.63).

Step Two: Determining the Altitude of the Measuring Triangle

With the base of the measuring triangle in hand, it was easy to determine both the altitude and the hypotenuse. The altitude of this measuring triangle was, in fact, too easy to be useful as an example. Because we wanted to build a roof with a 12-in-12 pitch, the altitude had to be the same number as the base, or 134.63 in.

Let's pretend for a moment that I wanted a slightly steeper roof, one that had a 14-in-12 pitch. In a 14-in-12 roof, there are 14 in. of altitude for every 12 in. of base. To get the altitude of the measuring triangle, then, I would find out how many 12-in. increments there are in the base, then multiply that number by 14. In other words, divide 134.63 by 12, then multiply the result by 14. Here's what the math would look like: 134.63 in.÷12 = 11.22; 11.22 x 14 = 157.08 in. Finding the hypotenuse of the measuring triangle is just as simple.

Step Three: Determining the Hypotenuse of the Measuring Triangle

Now let's return to our 12-in-12 roof. The base and the altitude of the measuring triangle are both 134.63 in. But what's the hypotenuse? One way to solve the problem of finding the hypotenuse is to use the Pythagorean theorem: $A^2 + B^2 = C^2$ (where A and B are the legs of the triangle and C is the hypotenuse).

There are other ways to solve this problem—with a construction calculator, with rafter manuals, with trigonometry—but I usually use the principle mentioned in step one. According to this principle, you can expand a triangle without changing the angles by multiplying all three sides by the same number.

For over a century, carpenters have used the rafter tables stamped on rafter squares. The common table shows the basic proportions of triangles for 17 different pitches.

The base of all these triangles is 12; the altitude is represented by the number in the inch scale above the table. And the hypotenuse is the entry in the table. Under the number 12, for example, the entry is 16.97. This is the hypotenuse of a right triangle with a base and an altitude of 12.

To use this information to create the larger measuring triangle I needed for this roof, I simply multiplied the altitude and the hypotenuse by 11.22. This, you may recall, was the number I obtained in step two when I divided the base, 134.63 in., by 12. Now multiply the altitude and the hypotenuse of the small triangle by 11.22: 11.22 x 12 gives us an altitude of 134.63 in.; and 11.22 x 16.97 gives us a hypotenuse of 190.40 in.

So here is the technique I use for any gable roof. I find the base of the measuring triangle, divide it by 12, and multiply the result by the rise of the pitch to get the altitude of the measuring triangle (which determines the height to the bottom of the ridge). To get the hypotenuse, I divide the base of the triangle by 12 and multiply that by the hypotenuse of the pitch, which is found in the common rafter table.

Rafter Square Gives You the Hypotenuse

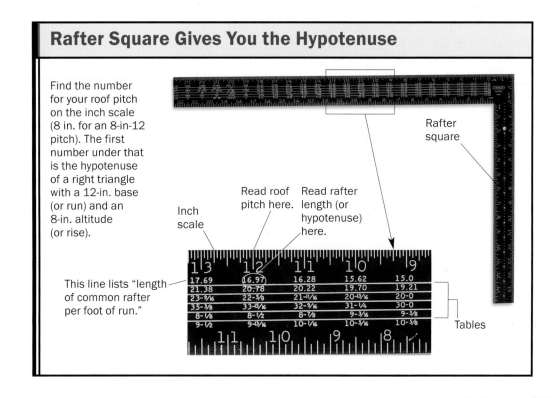

Find the number for your roof pitch on the inch scale (8 in. for an 8-in-12 pitch). The first number under that is the hypotenuse of a right triangle with a 12-in. base (or run) and an 8-in. altitude (or rise).

This line lists "length of common rafter per foot of run."

Inch scale

Read roof pitch here.

Read rafter length (or hypotenuse) here.

Rafter square

Tables

	13	12	11	10	9
	17.69	16.97	16.28	15.62	15.0
	21.38	20.78	20.22	19.70	19.21
	23-3/8	22-3/8	21-11/16	20-13/16	20-0
	35-3/8	33-13/16	32-3/8	31-1/4	30-0
	8-1/8	8-1/2	8-7/8	9-3/16	9-3/8
	9-1/2	9-13/16	10-1/16	10-3/16	10-3/8

Say the roof has an 8-in-12 pitch with a base of 134.63. I divide that number by 12 to get 11.22, then multiply that by 8 to get the ridge height of 89.76 in. To get the length of the rafter, I multiply 11.22 by 14.42 (the number found under the 8-in. notation on the rafter table), for a length between ridge and bearing wall of 161.79 in.

The only time I waver from this routine is when a bird's mouth cut the full depth of the wall leaves too little wood to support the eaves. How little is too little depends on the width of the rafter and the depth of the eave overhang, but I generally like to have at least 3 in. of uncut rafter running over the bird's mouth. If I have too little wood, I let the bottom edge of the rafter land on top of the wall rather than aligning with the wall's inside edge. Then I determine how far the rafter will sit out from the inside edge of the bearing wall and use that inside point as the start of my measuring triangle.

Step Four: Setting the Ridge

I determined that the altitude of the measuring triangle was 134.63 in. This meant that the correct height to the bottom of the ridge was 134⅝ in. above the top plate of the wall. (Note: I usually hold the ridge board flush to the bottom of the rafter's plumb cut rather than the top.) To set the ridge at this height, we cut two posts, centered them between

Setting the ridge. Temporary posts are set up to hold the ridge at the right height. The posts are braced 2x4s with a 2x4 scrap nailed to the top that rises 10 in. above the top of the post. The ridge is set on top of the post and nailed to the scrap of wood.

the bearing walls, and braced them plumb. Before we installed the posts, we fastened scraps of wood to them that ran about 10 in. above their tops. Then, when we set the ridge on top of the posts (photo, facing page), we nailed through the scrap into the ridge. We placed one post against the existing house and the other about 10 in. inside the gable end of the addition. This kept it out of the way later when I framed the gable wall.

Getting the ridge perfectly centered is not as important as getting it the right height. If opposing rafters are cut the same length and installed identically, they will center the ridge.

Step Five: Laying Out the Main Part of the Rafter

I calculated a measuring length (or hypotenuse) for the rafter of 190.40 in., then used the jig as a cutting guide and made the plumb cut. I measured 190⅜ in. (converted from 190.40 in.) from the heel of the plumb cut and marked along the bottom of the rafter. Rather than have another carpenter hold the end of the tape, I usually clamp a square across the heel of the plumb cut, then pull the tape from the edge of the square (photo, right) to determine where the bird's mouth will be. Once I marked where the bird's mouth would be, I used the jig to mark a level line out from the mark. This would be the heel cut, or the portion of the rafter that sits on top of the bearing wall.

After scribing a level line, I measured in the thickness of the wall, which was 5½ in., and marked. I slid the jig into place and scribed along the plumb edge from the mark to the bottom edge of the rafter. This completed the bird's mouth and, thus, the layout for the main portion of the rafter.

A clamped square serves as an extra pair of hands. It's easy to measure along the bottom of the rafter if you clamp a square to the bottom of the plumb cut and run your tape from there.

Visualizing the bird's mouth and rafter tail. Red lines on this marked-up rafter show where cuts will be made for the bird's mouth, at right, and rafter tail, at left. To determine the correct cut for the tail, the author marked out the subfascia, soffit, and fascia board.

Step Seven: Preserving the Layout

The only dimension for this layout that I had to remember was 190⅜ in., the hypotenuse of the triangle and the measuring length of the rafter. I wrote the number where I could see it as I worked. To preserve the other three critical dimensions—one for the plumb cut of the bird's mouth and the other two for the rafter tail—I used the rafter jig to extend reference points to the bottom edge of the rafter; then I transferred these marks to a strip of wood, or measuring stick (top photo, facing page). I was ready to begin cutting the rafters.

Step Eight: Marking and Cutting the Rafters

Some carpenters lay out and cut one rafter, then use it as a pattern for the rest, and I'll often do that on a smaller roof. On this roof, where the rafters were made of 20-ft.-long 2x10s and where I was laying them out by myself, this method would have meant a lot of heavy, awkward, unnecessary work. Instead of using a 100-lb. rafter as a template, I used my jig, my tape measure, and the measuring stick.

Moving to the end of the 2x12, I clamped the jig in place and made the plumb cut. (The steep pitch of this roof made clamping the jig a good idea. Usually, I just hold it to the rafter's edge the way you would when using a framing square as a cutting guide.) Then I clamped my square across the heel of that cut, pulled a 190⅜-in. measurement from that point and marked along the bottom edge of the board. Next I aligned the first mark on the measuring stick with the 190⅜-in. mark and transferred the other three marks on the measuring stick to the bottom edge of the rafter (top photo, facing page).

Step Six: Laying Out the Rafter Tail

The eaves on the existing house measure 16 in. out from the exterior wall, which meant I would make the eaves on the addition 16 in. wide. To lay out the rafter tail, I started with the finished dimension of 16 in. and then drew in the parts of the structure as I envisioned it (photo, above). In this way, I worked my way back to the correct rafter-tail layout.

I began by holding the jig even with the plumb line of the bird's mouth. With the jig in this position, I measured and marked 16 in. in from the corner along the level edge. Then I slid the jig down to this mark and scribed a vertical line. This line represented the outside of the fascia. Next I drew in a 1x6 fascia and a 2x subfascia. I also drew in the ⅜-in. soffit I would use. This showed me where to make the level line on the bottom of the rafter tail.

Marking the bird's mouth and tail. Once the start of the heel cut was determined by measuring from the bottom of the plumb cut, this measuring stick was used to transfer the dimensions of the bird's mouth and rafter tail. The cutting lines were marked using the rafter jig.

To finish the layout, I used the jig to mark the level and plumb lines of the bird's mouth and the rafter tail (photo, right). For all four of these lines, I kept the jig in the same position and simply slid it up or down the rafter until either the plumb or level edge engaged the reference mark. It was quick and easy. It was fun.

The two cuts that formed the rafter tail were simple, straight cuts that I made with my circular saw. To cut the bird's mouth, I cut as far as I could with my circular saw without overcutting, then finished the cut with my jigsaw.

John Carroll is a builder in Durham, North Carolina. This article is adapted from his book, Measuring, Marking and Layout: A Guide for Builders, *published by The Taunton Press, Inc.*

A rafter jig does the hard work. Once the location of the bird's mouth is determined, the rafter jig is used to mark the level and plumb cuts of the bird's mouth and rafter tail.

A Gable-Dormer Retrofit

■ BY SCOTT McBRIDE

Carol and Scott Little's home draws its inspiration from the cottages of colonial Williamsburg and the one-and-a-half-story homes of Cape Cod. Both styles typically feature a pair of front-facing gable dormers. But for some reason, the builder of the Littles's house put only one dormer on the front, leaving the facade looking unbalanced. I was hired to add a new dormer on the front of the house to match the existing one and, for more light, added a scaled-down version of the same dormer on the back of the single-story wing.

As the crew set up the scaffolding and rigged the tarps against the possibility of rain (sidebar, p. 80), I crawled under the eaves to study the existing roof. I soon realized that framing the sidewalls of the two dormers and directing their load paths would require different strategies, as would the way the dormer ridges would be tied to the main roof.

The first consideration in a retrofit is the location of the dormers, and the second is their framing. The existing front dormer fit neatly into three bays of the 16-in. o.c. main-roof rafters. These main-roof rafters (or commons) were doubled up on each side of the dormer, creating the trimmer rafters that carry the roof load for the dormers. Full-height dormer sidewalls stood just inside these trimmers, extending into the house as far as the bedroom kneewalls. Additional in-fill framing completed the dormer walls that were above the sloped bedroom ceiling.

Fortunately, three rafter bays at the other end of the roof landed within a few inches of balancing with the location of the existing dormer. Consequently, I had only to sister new rafters to the insides of the existing ones to form the new trimmers, and I could match the framing of the existing front dormer, leaving a uniform roof placement, appearance, and size.

Cut the Opening and Shore Up the Main-roof Framing First

After laying out the plan of the front dormer on the subfloor, I used a plumb bob to project its two front corners up to the underside of the roof sheathing. Drilling through the roof at this location established the reference points for removing the shingles and cutting the openings.

Framing around an existing roof required different skeletons for each dormer.

The tricky part was establishing how far up the slope to cut the opening. To play it safe, I first opened just enough room to raise the full-height portion of the sidewalls (drawing, p. 81). With those walls up and later with some dormer rafters in place, I could project back to the roof to define the valley and then enlarge the opening accordingly.

Inserting new rafters into an already-sheathed roof can be problematic because of the shape of the rafters. They are much longer along the top edge than along the bottom, so there's no way to slip them up from below. A standard 16-in. bay doesn't afford nearly enough room to angle them in, either. To form the new trimmer rafters,

we cut the new members about 6 in. short of the wall plate before we secured them to the existing rafters.

When faced with this situation, I normally use posts to transfer the load from the trimmer rafters to an above-floor header. In fact, I did follow this step with the smaller rear dormer (sidebar, p. 82), but that would not work in this case. Here, the floor joists ran parallel to the front wall, instead of perpendicular to it, and so could not transfer the load to the wall.

I decided simply to let the existing single rafter carry the load for the last 6 in. to the front wall plate. This situation is not the ideal solution, but the weight of the dormer is not great enough to overtax the rafters over

Protecting the Roof

To protect the exposed roof against rain, we rolled up new poly tarps around 2x4s and mounted them on the roof ridge above each dormer. The tarps were rolled down like window shades each evening, with some additional lumber laid on top as ballast. The ballast boards were tacked together as a crude framework so that they would not blow away individually in high winds.

Supporting the Dormer

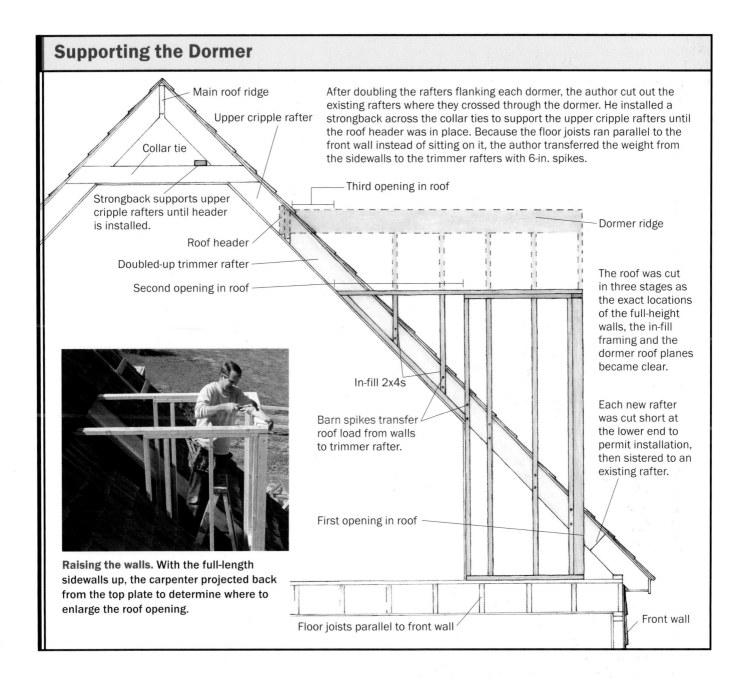

Main roof ridge

Upper cripple rafter

Collar tie

Strongback supports upper cripple rafters until header is installed.

Roof header

Doubled-up trimmer rafter

Second opening in roof

After doubling the rafters flanking each dormer, the author cut out the existing rafters where they crossed through the dormer. He installed a strongback across the collar ties to support the upper cripple rafters until the roof header was in place. Because the floor joists ran parallel to the front wall instead of sitting on it, the author transferred the weight from the sidewalls to the trimmer rafters with 6-in. spikes.

Third opening in roof

Dormer ridge

The roof was cut in three stages as the exact locations of the full-height walls, the in-fill framing and the dormer roof planes became clear.

Each new rafter was cut short at the lower end to permit installation, then sistered to an existing rafter.

In-fill 2x4s

Barn spikes transfer roof load from walls to trimmer rafter.

First opening in roof

Raising the walls. With the full-length sidewalls up, the carpenter projected back from the top plate to determine where to enlarge the roof opening.

Floor joists parallel to front wall

Front wall

such a short span, and doubling up the new trimmer rafters would at least stiffen the existing rafters considerably.

With the new trimmer rafters in, the existing main-roof rafters falling between them were cut and partially torn out to make room for the dormers. The portions above and below the dormer would remain as cripple rafters. The lower cripple rafters were plumb-cut in line with the dormer front wall where they would be spiked to cripple studs. Rough cuts were then made at the top, leaving the upper cripple rafters

long. These rafters would be trimmed back further only later, after we established the precise location of the dormer-roof header.

To support the upper cripple rafters temporarily, I climbed up into the little attic above the bedroom. There I laid a 2x4 strongback across four collar ties, including the collar ties connected to the recently doubled trimmers. This strongback would support both the collar ties and the upper cripple rafters until we could install the dormer-roof header.

Back Dormer Demands Different Strategies

Unfortunately, when it came to the smaller dormer in back, the existing rafter layout did not match where the dormer needed to be, as was the case in front. Here, I had to build new trimmer rafters in the middle of the existing rafter bays.

The attic space differed, too. Whereas the front dormer served a bedroom, the back dormer was in a storage room. Because the owner wanted to maximize floor space in this storage area, I built the sidewalls on top of the rafters, which pushed the kneewall back and allowed the ceiling slope to extend all the way to the dormer's gable wall (drawing, below).

As in the front dormer, we cut the new trimmer rafters short. This time, however, the floor joists ran perpendicular to the front and back walls, which meant that I could use posts to transfer the load from the trimmer rafters to an above-floor header (photo, facing page). The header

Above-Floor Header Distributes Weight of the Dormer

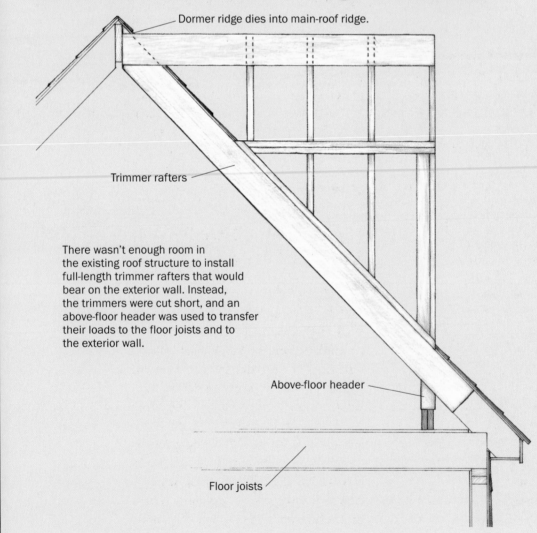

Dormer ridge dies into main-roof ridge.

Trimmer rafters

There wasn't enough room in the existing roof structure to install full-length trimmer rafters that would bear on the exterior wall. Instead, the trimmers were cut short, and an above-floor header was used to transfer their loads to the floor joists and to the exterior wall.

Above-floor header

Floor joists

Spreading the load. A doubled-up 2x6 sits on top of the attic floor and spreads the load of the new dormer over four floor joists.

distributes the weight over several floor joists, and the joists carry the weight back to the wall. The additional strain imposed on the floor joists is minimal because the header is so close to the wall.

The location of the back dormer's roof ridge altered another aspect of the framing. Because this dormer's roof ridge was at the same elevation as the main-roof ridge, I tied the dormer ridge and the main-roof ridge directly together instead of building a separate dormer header.

Measuring and cutting the valley rafters was the same for the back as for the front with the exception that at the bottom of the back dormer's valley rafters, the compound miters did not need a level seat cut because the valley rafters would not sit on top of 2x4 wall plates. Instead, the valley rafters were simply nailed to the face of the trimmer rafters.

Dormer Sidewalls Can Be Framed Two Ways

With trimmer rafters installed and cripple rafters secured, I could proceed with the walls. I know two common ways to frame dormer sidewalls: You can stand a full-height wall next to a trimmer rafter, or you can build a triangular sidewall on top of a trimmer rafter, which is how I framed the rear dormer. To match the new front dormer to the existing one, I used full-height studs 16 in. o. c. only as far in as the kneewall.

This type of dormer sidewall normally delivers the weight of a dormer to the floor. In new construction, this weight is taken up by doubling the floor joists under these walls. I didn't want to tear out the finished floor, however, so I joined the full-length sidewalls to the trimmer rafters by predrilling and pounding two 6-in. barn spikes through each stud. This transferred the load to the trimmer rafters rather than placing it on the floor framing. I've seen barn spikes withstand tremendous shear loads in agricultural buildings, so I felt confident they could carry this little dormer.

Plan the Cornice Details before Framing the Roof

With the walls up, the roof framing, which is the most complicated, came next. Before cutting any dormer rafters, though, I drew a full-scale cornice section, using the existing dormer as a model. Worrying about trim before there is even a roof may seem like the tail wagging the dog; but it makes sense, especially in a retrofit. The existing dormer featured a pediment above the window. The eaves had neither soffit nor fascia, just a crown molding making the transition from the frieze board to the roof (photo, p. 87). That detail reduced the dormer rafter tail to

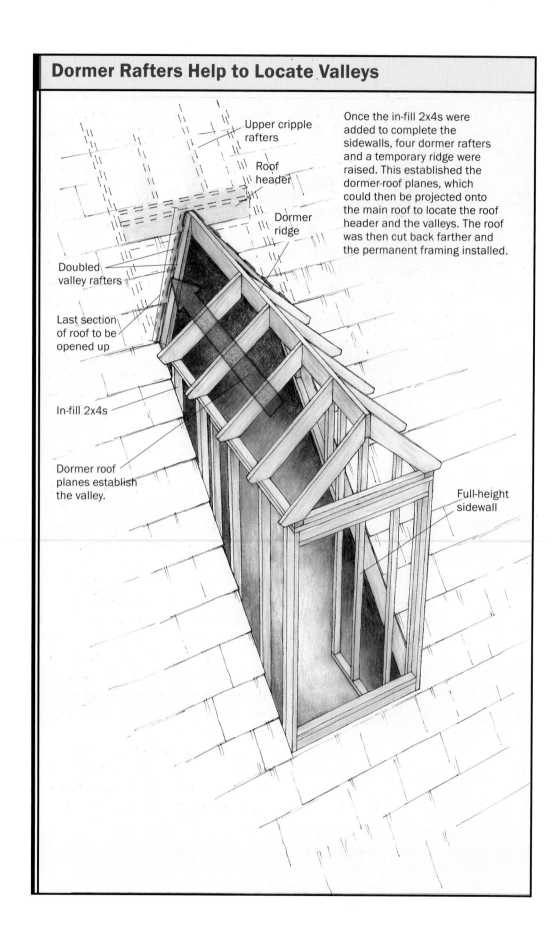

Dormer Rafters Help to Locate Valleys

Upper cripple rafters

Roof header

Dormer ridge

Doubled valley rafters

Last section of roof to be opened up

In-fill 2x4s

Dormer roof planes establish the valley.

Full-height sidewall

Once the in-fill 2x4s were added to complete the sidewalls, four dormer rafters and a temporary ridge were raised. This established the dormer-roof planes, which could then be projected onto the main roof to locate the roof header and the valleys. The roof was then cut back farther and the permanent framing installed.

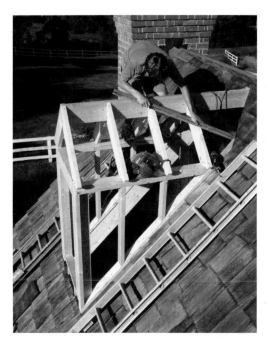

Finding the bottom of the valley. A straightedge is laid across the dormer rafters to project the roof surface to the inside of the trimmer rafter.

a mere horn that would catch the top of the crown molding. The eaves section drawing helped to establish the cuts for the rafter tails and trim details.

Along the rakes, the crown molding was picked up by the roof sheathing, which was beveled and extended out past the gable wall. Using a short piece of molding as a template, I worked out the amount of the overhang and the correct bevel for the edge of the sheathing in the rake-section drawing. Juxtaposing the drawings ensured that the rake crown, the eave crown and the level-return crown would all converge crisply at a single point.

Framing the Roof Defines the Valleys

Ready to proceed with the roof framing, we set up two pairs of common rafters with a temporary ridge board between them. Then we used a straightedge to project the outline of the dormer roof planes onto the main

roof and cut back the main-roof sheathing accordingly (drawing, facing page). Having established the elevation of the dormer ridge, we trimmed back the upper cripple rafters and then installed the roof header to carry the permanent dormer ridge board. The roof header spans between the trimmer rafters, carrying the dormer ridge and the valley rafters. (On the rear dormer, the ridge was level with the main-roof ridge, so no header was necessary there.)

When the dormer common rafters and ridge were installed permanently, we used the straightedge again to find the intersection of the dormer roof planes and the inside face of each trimmer rafter (photo, left). This point is where the centers of the valley rafters would meet the trimmer rafters. At their tops, the valley rafters would nuzzle into the right angle formed between the dormer ridge and the main-roof header.

I like to "back" my valley rafters, a process of beveling them so that they accept the sheathing of each adjoining roof on its respective plane. Because a cathedral ceiling was to wrap under the valley, I backed the lower edge of the valley as well, giving a nice surface for attaching drywall.

In addition to backing, I double valleys, even when not structurally necessary, because it gives ample bearing for plywood above and drywall below. Doubling valley rafters also simplifies the cheek-cut layout at the top and bottom of the valley because a single compound miter is made on each piece instead of a double compound miter on a single piece.

Because of the dormer's diminutive size, valley jack rafters weren't required. Consequently, with the valleys in place, the framing was complete, and we could dry it in.

Keep the water moving down. An apron flashing seals the front wall with its ends bent around the corners (top left photo). Then the lowest step flashings have their vertical fins bent over to cover the notches in the apron (top right photo).

Water-table flashing protects window and trim. The crown that forms the bottom of the pediment will go below the flashed water table and miter with an eave crown (seen poking out past the corner).

Careful Sheathing and Flashing Combat Wind and Water

We sheathed the front of each dormer with a single piece of plywood for maximum shear strength (bottom right photo). With so little wall area next to the windows, I was concerned that the dormer might rack in high winds. The small back dormer was especially worrisome because it had no full-length sidewalls to combat racking, but the single piece of plywood on its front stiffened the whole structure. We extended the roof sheathing past the gable wall and beveled it to receive the rake crown molding.

Flashing work began with an aluminum apron flashing at the bottom of the dormer front wall (top left photo). The downhill fin

In the doghouse. Narrow dormers are prone to racking. To stiffen this one, the author sheathed the front wall with a single sheet of plywood.

of this flashing extends a few inches beyond both sides of the dormer, and its vertical fin was notched and folded back along the sidewall. Then the first piece of step flashing had its vertical fin folded back along the front wall to protect the corners where the apron flashing had been notched (top right photo, above). Step flashings march up along both sides of the dormer, with the uppermost pieces trimmed to fit tightly beneath the dormer roof sheathing. It was tough work weaving step flashings into the existing cedar-shake roof. Hidden nails had to be extracted with a shingle ripper, a tool with a flat, hooked blade. If I had it to do over, I would sever these nails with a reciprocating saw before the dormer sidewalls were framed.

The valley flashing was trimmed flush with the dormer ridge on one side of the roof, and the opposing valley flashing was notched so that it could be bent over the ridge. We protected the point where the valleys converge at the dormer ridge with a small flap of aluminum with its corners bent into the valley. This approach is more reliable than caulk.

The last piece of flashing to go on was the gable water-table flashing (bottom left photo, facing page). Its front edge turns down over the return crown molding, and its rear corners fold up under the extended roof sheathing to repel wind-driven rain.

Finish Trim Improves on Weather Performance of Existing Dormer

The house is just a few years old, but the existing front dormer had suffered extensive decay. In the worst shape were the finger-jointed casings and sill extensions that the original builder had used. To avoid a repeat of this calamity, I used only solid moldings and bought cedar for the trim boards. Everything was primed, especially the ends. To promote air circulation, the ends of corner boards and rake boards were elevated an inch or so above nearby flashings.

We wanted the new cedar shakes on the dormer to blend in with the existing weathered roof. The best solution we found was to brush on an undercoat of Minwax Jacobean®, followed by a top coat of oil-based exterior stain in a driftwood-type shade. The undercoat added a nice depth to the gray top coat.

Contributing editor Scott McBride is a carpenter and architectural woodworker in Sperryville, Virginia.

A crowning moment. Three pieces of crown converge at the bottom corners of the pediment. Trim, casings, and sills primed on every side, resist rot.

Supporting a Cantilevered Bay

■ BY MIKE GUERTIN

When a client wants to add curb appeal to a new home, I dip into my Mr. Potato Head® bag of tricks: a distinctive window here, a reverse gable there, fancy trim details, an entry portico or a porch—and voilà! It's enough to make an architect cringe.

One of the best-selling upgrades is an angled bump-out or bay. It adds a few square feet, creates a distinctive room inside, and dresses up the home's exterior. Although I could just install a bay window for light and effect, I find the floor-to-ceiling bay more appealing as well as competitive in cost.

But a bay is only as strong as the floor that it's built on. In this article, I'm going to concentrate on the proper techniques for framing the cantilevered floor that supports a bay. For this project, the bay was 8 ft. wide and extended 2 ft. from the house. The sides of the bay were set at 45°, but they could have been set at any angle.

Cantilevered Joists Save Foundation Work

Cantilevering the bay keeps down the cost, about $400 less than an angled foundation. It's also easiest to frame one of these bays when the joists run parallel to the floor framing. In this scenario, the common joists are just lengthened to form the bay joists, eliminating the need for headers and hangers. But I wasn't so lucky on this project. The floor joists of this bay ran perpendicular to the main joists (photo, on p. 90).

The cantilever wouldn't be carrying any loads but the bay itself, so I followed the two-thirds in, one-third out cantilever rule of thumb. With a 2-ft. cantilever, the bay joists would be anchored to a tripled floor joist 4 ft. in from the outside of the house. But I waited to add the second and third joists until just before sheathing the deck. Having only one common joist allowed me to nail through it to attach the bay joists initially. The bay joists follow the 16-in. o.c. layout regardless of exactly where the bay is placed, so first I put in all the 4-ft. joists that fell on each side of the bay area.

Plates before joists. Before the cantilevered-bay joists are cut and installed, the wall plates are cut and laid out. At this point, all measurements are checked, and the exact width of the bay is established.

Cut and Lay Out the Bay Plates First

Before I laid out the exact location of the bay, I cut and laid out the top and bottom plates for the bay walls (photo, below). Although this step may seem a bit premature, I always want to be certain that the windows will fit and that I'll still have room inside and out for the trim. The plates also help me to figure the length and cut for each joist.

A little basic math and a calculator gave me the plate lengths. With the bay cantilevering 2 ft. and the walls at 45°, I needed to come in 2 ft. from each side for the bay's front plate. With 22½° angles on each end (half of 45°), I cut the plate for the bay's front wall at 4 ft. from long point to long point.

With some help from Pythagoras, I cut the side plates again with 22½° angles on each end and with the outside face measuring 33¹⁵⁄₁₆ in. long point to short point (short point because the adjoining wall plate is also cut at 22½° to form the inside corner). With

the plates laid out on a flat surface, I marked the rough opening for the window centered on the 4-ft. plate.

To get the width of the trim (exterior and interior) to match on both sides of the bay's outside corners, I make sure that the distance is the same from the corners to the edge of the rough openings for all three windows. After the rough openings are marked out, I also make sure that I have enough space left (at least 1 in.) for the inside-corner trim.

Center the Bay on the Interior

I usually center a bay on the room inside. In this case, that threw it slightly off-center on the exterior, but the difference wouldn't be noticeable. I marked the location of the 8-ft. opening on top of and on the outside face of the sill plate. (On this house, the sill plate is actually the top plate of a framed wall for a walk-out basement.)

Next, I marked the outside corners of the bay on the sill, showing me which joists would cantilever the full distance. The

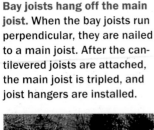

Bay joists hang off the main joist. When the bay joists run perpendicular, they are nailed to a main joist. After the cantilevered joists are attached, the main joist is tripled, and joist hangers are installed.

Setting the overhang. With a 2-ft. cantilever, the longest joists overhang 22½ in. The joists are then tacked in place.

Bay starts here. After the bay joists are tacked in place, the outside edge of the bay is carried up from the face of the sill onto the rim joist.

house's rim joists were then run to the locations of the first cantilevered joists inside the 8-ft. layout marks rather than being mitered into the bay's rim joists. The extended rim joists are nailed square to the cantilevered joists to hold them straight and plumb, and in turn, they provide a more solid place to secure and plumb the bay's rim joists where they meet the house wall.

The cantilevered bay joists were put in next. The joists that fell in the middle 4-ft. section of the bay cantilevered by 22½ in. (2 ft. less the thickness of the rim) (top left photo). These joists were tacked to the sill and nailed to the common joist. The joists that fell on the angled sidewalls were left a little long and cut to exact length later.

When all the joists were nailed in, I ran a string to straighten the main rim joist and then drew square lines up from the lines I'd made earlier on the sills, indicating the outside edges of the bay (top right photo). The rim joist for the outer wall of the bay was then cut and nailed in, left long to be cut to exact length later (photo, right).

Outer rim joist is left long. Rim-joist stock is nailed to the cantilevered joists and left long until cutlines are transferred from the plates.

Precut wall plates provide the shape of the bay. The bay plates are laid on top of floor joists and rim joists that were run long and needed to be cut.

Cutlines are squared down from the plates. Lines are extended down from the corners of the plates for the rim-joist cuts, and the lengths of cantilevered side joists are marked 1½ in. from the outside edge of the plate.

Cut down where they stood. The cantilevered side joists and the outside rim joist are cut to length in place.

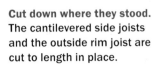

Use Plates to Mark the Joist Cuts

With the joists in place, I next set the bay's wall plates in position over the cantilevered joists and rim (top photo, facing page). At the outside corners of the plates, I squared down a cutline indicating where to miter the bay's rim joist (bottom left photo, facing page).

A line was also drawn along the outside edge of the sidewall plates onto the tops of the joists that were left long. This line is the perimeter of the bay, so holding a 2x block inside the line and drawing a second line allowed for the rim joist and gave me the actual cutline. After squaring down the cutlines, the joists were trimmed at 45° bevels (bottom right photo, facing page). If the amount of waste is more than about a foot, I rough-cut the length so that the cutoff isn't heavy and unwieldy. Finally, I cut, fit, and installed the angled rim joists (top photo).

With all the floor framing for the bay complete, I tripled the main joist that the bay hung from and installed two joist hangers on each of the cantilevered joists, one right side up and the other upside down (middle photo). The theory is that the upside-down hanger prevents the joist from lifting from the weight of the cantilever. Some framing crews block between the cantilevered joists at the sill plate, but I prefer to leave the space open to slide in insulation later.

With floor framing done, I finished sheathing the main deck, extending the sheathing over the cantilever (bottom photo). I left the walls and roof of the bay until the entire house was framed. The house walls give me a nice surface for attaching the bay roof, and a roomy exterior deck wrapped around the bay, disguising the fact that it was sitting on a sturdy cantilevered floor framing.

Mike Guertin is a custom homebuilder and remodeler in East Greenwich, Rhode Island. He is a contributing editor for Fine Homebuilding *magazine, the author of* Roofing with Asphalt Shingles, *and coauthor of* Precision Framing, *all published by The Taunton Press, Inc.*

Interlocking joist connection. The main rim joist on the house was run to the first cantilevered joist and nailed, keeping both joists plumb and square, and providing a solid landing place for the angled rim joist.

Wait—those joist hangers are upside down! To prevent uplift, joist hangers are nailed on top of the bay joists as well as on the bottom, where they support the weight of the floor.

Sheathing completes the floor framing. With the bay joists on regular centers, the deck sheathing can be cut and extended to include the bay without changing the sheathing pattern.

Remodeling with Metal Studs

BY TOM O'BRIEN

I recently went into a lumberyard to pick up a few 2x4s. The bill left me wondering if maybe I should have stopped off at the bank for a second mortgage. Sadly, anyone who has purchased lumber lately has probably had a similar experience. But what alternative is there to the high price and dubious quality of framing lumber? For more than 10 years metal studs have been the answer for me. In Virginia, lightweight 25-ga. metal studs for remodeling are a little more than half the cost of good-quality 2x4s.

I don't suggest abandoning wood completely in favor of steel. Despite cost advantages, metal framing does have drawbacks that limit its effectiveness for total residential framing. Load-bearing partitions require a more costly, heavier-gauge steel that has to be cut with special tools and must be welded or fastened with expensive drill-tipped screws. Also, the thermal conductivity of steel makes insulating a steel-studded wall more difficult. For these reasons I still choose wood for framing exterior and load-bearing walls. But for other framing applications, I find light-gauge metal faster, cheaper, and easier to work with than wood.

Try carrying ten 2x4s like this. Steel framing weighs much less than wood, and studs come in interlocking bundles of ten, which makes them easier to carry.

Steel Framing Is Stable and Uniform

Until lumber prices went out of sight, material costs for wood and steel were roughly the same. I was using metal framing then because my labor costs were lower. The reasons are simple. Metal framing is a manufactured product, which means that it's stable, straight, and uniform. These qualities translate into time saved that would be wasted digging through piles of wood-framing stock, looking for acceptable material, and sorting and crowning at the job site. Product

stability also eliminates the need to repair or replace metal-framing members that warp or distort after they have been installed.

Metal framing is easier to handle than wood because it weighs significantly less. Studs, for instance, come in easy-to-carry, interlocking bundles of 10 (photo, above). Steel boasts other advantages over wood, including resistance to damage by fire, insects, and weather. Steel framing is stocked in standard sizes from 1⅝-in. to 6-in. widths up to 20-ft. lengths.

Tools for steel framing should look familiar. All the tools you'll need for framing with steel are probably already in your toolbox. They include spring clamps, a 2-ft. level, a chalkline, a plumb bob, tin snips and a cordless drill. A screw gun (background) and locking C-clamp pliers are also helpful.

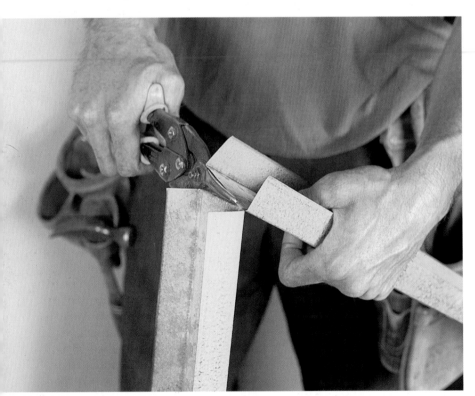

Cutting lightweight steel is quick and easy with tin snips. After cutting through the flanges of the track, or stud, bend the waste end back and cut through the web with a slightly circular inward motion.

Lightweight Steel Framing Requires No Special Tools

Chances are most of the tools needed for metal framing are already in your toolbox. These tools include a measuring tape, square, plumb bob, level (a magnetic level is handy but not necessary), chalkline, and tin snips. For most jobs, all that's needed to cut metal studs is ordinary, straight-cutting tin snips. Metal-cutting chopsaws that can slice through entire bundles of studs at one time are available for about $200. However, I've gotten along just fine all these years using tin snips to cut steel framing. Here's how I do it.

First, I mark the stud the same way I would a 2x4 except that I use a felt-tip marking pen (or a grease pencil) because it shows up better on metal than pencil does. I cut through the flanges on both sides of the stud with my tin snips, then turn the stud over and bend the cut end back. I then cut through the back of the stud (the web) with a slightly circular inward motion (photo, left). If you are unable to cut all the way through from one direction, just turn the stud around and finish the cut from the other side.

A VSR Screw Gun Facilitates Assembly

For fastening metal framing as well as attaching drywall, the one substantial tool purchase I would recommend is a variable-speed reversing (VSR) screw gun. This tool streamlines the assembly process. I prefer a 0–2,500 rpm model because it is geared lower for more power and better control, but a 0–4,000 rpm model works fine and is usually less expensive.

Another tool that makes framing with metal much easier is 4-in. locking C-clamp pliers made by Vise-Grip®. A pair of these is well worth the small investment. I also keep a couple of 2-in. spring clamps handy for suspending my plumb bob from the ceiling-track flange.

Self-Drilling Screws Connect Steel Framing Members

Screws are the fastest and easiest way to fasten metal framing. Framing members are usually joined with ⁷⁄₁₆-in. type-S self-drilling pan-head screws, commonly referred to as framing screws. Avoid type-S12 screws with drill-point tips, which are designed for fastening heavy-gauge steel. These screws are more expensive and tend to strip out light-gauge metal.

To join framing members, I first clamp the two pieces in place with my Vise-Grips (photo, right). Next I put a screw on the magnetic bit in my screw gun and hold it against the track, keeping the gun as perpendicular to the track as possible. I start the gun as if drilling a hole, and when the screw begins to penetrate, I back off slightly on speed and pressure and let the screw thread itself home. I advise using a professional-quality #2 magnetic Phillips bit and changing the bit at the first sign of wear.

Drywall is attached to metal studs with type-S drywall screws, which have finer threads and sharper points than the type-W screws used with lumber. I use 1⅛-in. screws for single-layer drywall and 1⅝-in. screws where two layers are specified. I don't know anyone who still uses nails to install drywall, so hanging it on metal should not be a difficult transition; just remember to use a lighter touch on the screw gun.

Casing and baseboard can be nailed to wood blocking installed in the metal framing, but it is simpler to attach trim with finish screws, which look and work the same as finish nails. They have small, self-countersinking heads, which are designed to be driven below the surface of the wood and covered with putty. These days, most finish screws have square-drive heads that work better than Phillips heads but still require a bit of care to avoid strip outs. Again, these screw tips should be changed at the first sign of wear. The lengths of trim screws most commonly used are 1⅝ in. and 2¼ in. The shorter ones are easier to work with but are just barely long enough to apply ¾-in. trim over ½-in. drywall. I try to keep some of the longer screws handy at all times.

It is possible to attach softwood trim to a metal-framed wall without predrilling, but you're probably asking for trouble. If you don't predrill, screws may bend, creating large, ugly holes as they go in, or they may refuse to countersink. It's also difficult to keep the wood tight to the wall and to keep it from splitting. I usually keep a ⅛-in. bit chucked in my cordless drill and alternate between that and my screw gun, although I

TIP

Use a professional-quality #2 magnetic Phillips bit when fastening metal framing and change the bit at the first sign of wear.

Clamping pliers aid assembly. Locking pliers with a C-clamp head keep the steel studs in place until screws are driven. Each stud gets one screw through each side, top, and bottom.

Sources

Gold Bond® Building Company (Div. of National Gypsum)
2001 Rexford Road
Charlotte, NC 28211
800-628-4662
www.gold-bond.com

Publishers of Gypsum Construction Guide, *a 163-page booklet that offers technical information for steel framing.*

Unimast, Inc.
4825 N. Scott St.
Ste. 300
Schiller Park, Il 60176
800-323-0746
www.unimast.com

Publishers of Steel Framing Technical Information Catalog *(#UN 30) and* Construction Worker's Guide to Steel Framing *(#UN 81).*

have drilled and fastened trim using only my cordless drill with a quick-change bit system. When predrilling, it's necessary to stop before the bit hits steel framing. The screw should be driven so that the head is countersunk about ⅛ in.

Basic Framing Process Is Similar to Wood

Except for a few simple differences, framing with metal is basically the same as stick framing with wood. Plates in steel framing are made of U-shaped channel, and the studs have a C-shaped profile (photo, p. 94).

To frame a basic metal-stud wall, I start by laying out the bottom plate as I would with any wall. After establishing the two end points of the wall, I snap a line on the floor and mark on which side of the line I want the wall. I transfer this line to the ceiling using a plumb bob or a level alongside a steel stud and always check to make sure the top plate is marked on the same side of the line as the bottom. Next, I cut the top and bottom plates out of track stock, screw them into place, and mark the stud layout on the tracks. At this point I take a few stud height measurements at various locations. If they differ by ½ in. or less, I subtract ⅛ in. from the shortest measurement and cut all studs to that length. Because the studs fit between the flanges of the track and are screwed in place, it is perfectly acceptable for stud height to vary by as much as ½ in.

After cutting the studs to length, I stand them up, setting the bottom of each stud inside the track lengthwise and tilting the top into its approximate position in the top track. Next, I twist the stud a quarter-turn so that the flanges of the track grip it and hold it in place.

All of the studs can be erected in this manner before they are fastened. After orienting the studs on the layout marks, I roughly split the gap in stud length between the top and bottom track and clamp the

bottoms. The bottom of each stud gets fastened with a screw driven through each side of the track (photo, p. 97). Finally, I climb a ladder and secure the tops.

A couple of steel-framing quirks need to be pointed out. First, most metal studs come with prepunched holes to accommodate plumbing and electrical systems. These holes need to be aligned before the studs are cut. Some of these punch-outs also have a definite top and bottom, which means that all the studs must be measured and cut from the same end.

Second, the open part of the steel studs should face the beginning of the stud layout. This placement lets the drywall contractors know where to start their sheets. More important, it prevents the studs from distorting when the drywall is attached, which keeps the seams between the sheets flat and even.

No Extra Studs Are Needed in the Corners

Framing with metal studs eliminates the need for special corner construction or extra framing because the drywall itself ties the intersecting walls together. When framing the corners of steel-framed walls, I first decide which wall will run by and which will butt just as in wood framing. I cut the track pieces for the bywall to the exact length and put them in position. Next, I measure and cut the track for the butt wall, leaving about a ¾-in. gap between the end of the track and the edge of the bywall track. This space is for the drywall. After cutting and positioning all the track, I double-check that each wall is laid out the same at the top and at the bottom and then anchor the track in place.

I lay out the studs for both the bywall and the butt wall, allowing for just a single stud at the end of each wall. Next, I cut and fasten all of the studs, except for the end stud of the butt wall. I leave this butt-wall stud floating until the drywall has been installed.

The drywall is an integral part of the corner. When building a corner of steel, the drywall on the inside is attached to the last stud of the bywall before the last stud of the adjoining wall is installed (left). That stud is fastened with screws driven through the drywall (below). This method simplifies construction and eliminates the need for extra framing in the corner.

After the blocking is in place and plumbing and electrical lines are roughed in, the walls are ready for drywall. I start inside the room on a bywall and run the board through the corner to the end of the wall (top photo, p. 99). Then I hold the floating corner stud from the adjacent butt wall against the drywall I just installed and fasten it with screws driven through the back of the drywall and into the web of the floating stud (bottom photo, p. 99). The top and bottom of the floating stud now can be screwed into the stud's track. This procedure ties the corner together and uses fewer studs than wood framing. The corner is completed structurally when drywall is applied to the outside of the walls.

Framing the intersection of two steel-studded walls is essentially the same as framing a corner. Again, no special framing is required. The track for the intersecting wall is kept short, and the end stud for that wall is back-screwed on after the drywall is installed on the main wall. Sometimes it's not possible to get behind the bywall to back-screw the intersecting stud. In these cases, I slide the last stud of the intersecting wall into position against the drywall and shoot pairs of 1⅛-in. drywall screws through the inside corners of the stud into the drywall at opposite 45° angles. A pair of these screws every foot or so holds the stud in place nicely until the drywall is installed.

To give a metal-studded wall additional rigidity, I stagger the seams of the drywall so that inside and outside sheets don't break on the same stud. Taping and finishing the drywall is the same as always: messy and tedious.

Door Framing Is Lined with Wood for Nailing

Framing for doors in a steel-studded wall requires extra effort. First, the studs must be at least 2½ in. wide to accommodate a door jamb. I start framing by locating the position of the door in the layout on the floor. I mark the centerline of the rough opening, and I measure half the rough-opening dimension in each direction.

I stop the track for the bottom wall plates at the edges of the rough door opening. However, I install the studs 1½ in. back from each side of the rough opening, which allows room to line the rough opening with 2x stock. This wood makes it easier to attach the door jamb and casing. I always check the door studs for plumb before screwing them in.

Metal tabs connect the door header to the studs. Door-header stock is cut long, the flanges are cut, and the ends are bent to create tabs that allow the header to be attached without special blocking.

To complete the door frame, I measure the rough-opening height off the high side of the floor, then add 1½ in. again for blocking. I mark this point on the stud with my marker and combination square, then use a level to determine the height of the other side. I cut the door header from a piece of track that's exactly 11 in. longer than the rough opening, and I square lines across the header 4 in. in from each end. This cut leaves me with the rough-opening width plus 3 in. for the blocking on both sides between my marks. I then make 45° angle cuts through the flanges of the track at both lines.

After making these cuts, I bend the 4-in. flaps down and install the header with the flanges facing up (photo, facing page). I clamp the tabs to the door frame at the correct height and screw through the tabs into the studs with two framing screws on each side. Finally, I continue my stud layout on the header and install the cripples between the header and the ceiling track.

I complete the door frame by attaching 2x blocking to the inside of the door-frame studs (photo, above). When I use 3⅝-in. studs, full-width 2x4s fit perfectly. But if I'm using 2½-in. studs to save floor space, I rip 2x6s down to about 2⅝ in. for the blocking. The blocking is then cut to length and attached to the steel studs with 1⅝-in. screws.

Steel Framing Is the Best Option for Hiding Masonry Walls

Here in Virginia, I do a lot of renovations of old brick row houses. Concealing unsightly brick or masonry walls has always presented a problem. These wall surfaces are rarely smooth enough to accept drywall directly, and using furring strips usually requires a lot

Drywall scraps help stiffen the wall. Pieces of drywall screwed to the studs with their edges against the wall behind help solidify a steel-studded wall that's only 1⅝ in. thick. The excess drywall is trimmed off later. Wood blocking is installed between studs to accept kitchen cabinets. The flanges of the steel studs have been flattened where the blocking passes by to keep the blocking flush with the stud face.

of fussing and shimming. Framing with 2x4s held away from the wall is fine but costly in space as well as money, and 2x2 walls seldom remain straight. My solution to this renovation problem is to use 1⅝-in. metal framing, which stays straight and requires little space.

First, I lay out the bottom plate 2⅛ in. off the brick (1⅝ in. for the stud width plus an extra ½ in. to allow for variations in the wall surface). After snapping the chalkline, I check a few points along the line to make sure the wall can be plumbed without hitting the brick. I also adjust the line for squareness to the room to accommodate any cabinetry, floor tiles, or intersecting walls. If necessary, the chalkline can be moved farther out.

When I'm satisfied with my layout, I use a plumb bob to transfer the layout to the ceiling for the top track. Next, I install both top and bottom track pieces. If I'm securing the track to wood floors or ceiling joists, 1⅛-in. drywall screws work fine. But if I'm attaching the track to a concrete floor, I use 1-in. powder-actuated fasteners. I finish framing the wall by cutting and installing the studs.

Compared with a 2x4 wall, an uncovered metal stud wall may seem downright flimsy. However, drywall applied correctly to both sides stiffens the wall. When I'm framing along a masonry wall, I can't apply drywall to the back of the studs. Instead, I install small scraps of drywall to act as stiffeners (photo, left). I hold each drywall scrap against the brick and screw it to the web of the stud, taking care not to bow the stud out into the room. I leave the edges of the scraps long and trim them off after they have been installed.

If kitchen cabinets are being installed, it's necessary to add solid blocking to a metal-framed wall (photo, left). This process can be done by first cutting up scrap 2x or ¾-in. plywood into 15¾-in. or 23¾-in. lengths, depending on the stud layout. Next, I mark all the pieces 1 in. from one end and cut a

¼-in.-deep saw kerf on each side of the line at that depth. This double kerf slips over the lip of the stud flange and allows the blocking to be installed flush with the outside of the studs. I often use another method of installing blocking without making saw kerfs. For that method, I bend the stud flange flat where the 2x passes by. In either case the blocking is attached with two 1⅛-in. drywall screws driven through the flange of the stud on one end of the block and through the inside of the adjacent stud on the other end.

After all the blocking is installed, the plumbing and electric have to be roughed in before the wall can be closed. I try to hire subcontractors who are familiar with metal framing because some procedures differ slightly from wood framing. Plumbers must be sure to isolate copper pipe from the steel framing with plastic bushings or tape to prevent galvanic corrosion. Electrical boxes may be screwed to the sides of the studs, fastened to wood blocking or attached with clips made specially for metal framing. Many electricians choose to run conduit through metal framing, but it's unnecessary as long as bushings are used to protect the wire sheathing from being damaged by the edges of the studs.

Before drywalling the outside of the walls, I sometimes insulate to provide thermal protection or soundproofing. Because metal studs are C-shaped and hollow, conventional insulation ends up being 1½ in. too narrow. Full-width insulation, 16 in. or 24 in., is available from your metal-stud supplier.

Framing Soffits with Steel Is a Breeze

When building interior soffits, I use metal framing almost exclusively. Whether it's a ceiling above kitchen cabinets, an enclosure for mechanical components, or a transition between rooms with ceilings of different heights, I can build a soffit straighter and more efficiently with 1⅛-in. metal framing than with wood (photo, below). Plus I don't need the arms of George Foreman to hold a section of lightweight steel framing over my head while driving screws.

For soffits, I use essentially the same framing procedure as I do with wood, although I prefer to build my soffits after the drywall has been applied to the adjacent walls and ceilings. This process makes layout easier, makes the soffit stronger, and eliminates the need for blocking (nailers) at the corners.

Tom O'Brien is a carpenter and remodeler in Richmond, Virginia.

Old technology meets new. A soffit is created of steel framing to conceal ductwork in the basement of an old house. The vertical members that will hold drywall are attached to sandblasted beams more than a century old.

Solo Drywall Hanging

■ BY PAT CARRASCO

As someone who hangs drywall for a living, I can say without a doubt that the first rule of solo drywall hanging is to get a good partner. Two people working together can get a lot more done, and with a lot less strain, than one poor slob stumbling around by himself. Having said that, I have to admit that even the best partners don't always show up for work. No matter how good the excuse—the first day of deer season, 2 ft. of fresh powder at Big Sky, "I woke up and felt so bad I figured it had to be Saturday"—for me, it's just another day spent holding my own. After 20 years of working in this trade, I've learned that many drywall projects can be done solo if I use the right tools and apply leverage rather than brute strength.

Real Men Can Use a Lift

In addition to the standard drywall tools that everyone uses—screw gun, router, utility knife, keyhole saw, 4-ft. square—a pair of step-up benches, a drywall foot pedal, and a soft hat are particularly useful when I'm forced to fly solo. I'll explain more about these tools, but first I need to mention the

one tool that's indispensable for hanging solo: a drywall lift. Basically a winch designed to raise and position large sheets of drywall safely, the drywall lift is available in most supply houses and rental shops. A few companies have made this tool, but the best one I've found is the PanelLift® from Telpro Inc.

I'm too cheap to purchase a drywall lift (they cost about $600), so when I need one, I rent (rental cost is about $20 a day). Tools like this take a lot of abuse, so when I stop at the local rental shop, I pick out the newest model in stock. I make sure the cable isn't frayed, and I give the wheels a good spin to see if the rubber and the bearings are in good condition. If the tool passes inspection, I pack it in my rig, and off I go. A drywall lift is a bulky, 100-lb. monster, but it's easy to transport because it breaks down into three compact sections that are easily reassembled on the job site.

Proper Preparation Keeps the Surgeon Away

Moving material from room to room is particularly difficult to do by myself, so if I know that I'll be working alone, I take

Use the right tools. A drywall lift makes it possible for one person to place large panels high up on walls and ceilings.

special care to make sure the drywall is stocked exactly where I want it. If the room is large enough for me to maneuver a full sheet—say at least 13 ft. by 13 ft.—I have the stockers put the board right in the room where I'll be working. Otherwise, the ideal location is in the hallway or a nearby room from which the board can easily be transported. When the delivery arrives, I ask the stockers to separate the sheets and stand them on edge against a wall with their white sides facing out (photo, p. 105). I make sure that the drywall is placed against the wall with enough of a tilt to prevent the pile from tipping over and injuring someone working on the job site. I feel safe if the bottom of the first sheet is 8 in. from the base of the wall.

I prep a room the same as I would for any drywall job, only more so: I check that all the framing surfaces are in the same plane and that all the nails are driven flush. I make sure that all the blocking is in place and that each corner has backing. I use a router to cut out for electrical boxes, so I tuck all the wires deep into their respective boxes. When I'm using a drywall lift, I sweep the floor thoroughly to make sure the wheels don't become caught on a stray screw as I'm trying to wrestle a 12-ft. sheet into position.

Large Sheets Go Up First

Professional drywallers usually hang the ceilings before the walls because we get extra wiggle room along the edges. To minimize the number of butt seams, we like to use large sheets wherever possible. Working with a partner, I may use sheets as long as 16 ft. When working alone, I limit myself to 12-ft. sheets. For efficient use of time and material, I put up all the large sheets first, then fill in the smaller pieces afterward.

Look, Ma! No hands! With the drywall lift doing the heavy work, eight 1¼-in. drywall screws are easily driven around the perimeter to tack the sheet.

Breaking between the joists. Rather than cut a factory edge to break on a joist, the author uses scrap sheathing material as a backer to create floating butt seams.

To load a 12-ft.-long sheet onto a drywall lift, the author bends his knees, places a hand under the corner, and straightens up, all the time keeping his back straight.

My first ceiling piece will be hung along one wall and perpendicular to the ceiling joists (photo, 105). If necessary, I adjust the height of my walk-up benches so that I can easily touch the ceiling while standing on the top step (left photo, facing page); then I place one or two benches alongside the wall and check the measurement for the first sheet. If I have to cut the sheet to make the seam break on a joist, I make the cut on the outside end, saving the factory edge for the butt seam. If I encounter an unusual framing configuration, I could also let the sheet break between the joists and use a backer to join the seams (right photo, facing page).

'Tilt and Lift' to Load Large Panels

The hardest part of solo drywall hanging is loading the board onto the lift without stress and strain. To give myself plenty of room to maneuver, I wheel the lift to the center of the room and lock the foot brake. Then I tilt the cradle body forward and unfold the support clips. Now all I have to do is persuade a 100-lb. slab of gypsum to get off the floor and set itself in the hooks.

The key to lifting a large sheet of drywall is a leveraging maneuver that I call the "tilt and lift." To make sure the board rests on the lift with the white (finish) side facing the frame, I pick up the sheet from the dark side. Standing at the end of the board, I squat down and reach a hand underneath the bottom corner. Without bending my back, I straighten up and raise the corner to waist height (photo, above). Ensuring that the sheet leans in toward my body, I reach the other hand down to the bottom edge and walk toward the middle of the sheet. The sheet will automatically tilt up to a horizontal position when my hands are centered. With the bottom edge balanced between my hands and the top leaning backward and resting against my shoulder, I walk over to the lift, center my hands between the clips and set the bottom edge on the clips (photo, p. 108). I then gently push the top of the sheet away from my body onto the cradle.

Once the board is settled in the cradle, the rest is easy. After tilting the cradle body backward so that the sheet lies flat, I fold the clips back and out of the way. Then I roll the lift to its approximate position in relation to the ceiling and crank the board up to

With the sheet balanced between his hands and leaning against his shoulder, the author carries it to the lift and deposits it in the support hooks.

within an inch of the ceiling (photo, 105). After fine-tuning the position, I crank the board up tight, reach for my screw gun and drive enough fasteners to hold it in place (left photo, p.106); a minimum of eight evenly distributed and correctly set screws is usually enough to support a 12-ft. sheet. Holding the crank with one hand, I release the crank brake with the other hand, ease the lift down and out of the way, then securely fasten the drywall with 1¼-in. type W screws, placed 7 in. o.c. along the butt seams and 12 in. o.c. everywhere else.

Small Sheets Go Up by Hand

Wherever I encounter a recessed light or an electrical box, I avoid the hassle of measuring and fitting by using a router to make the cut after the sheet is tacked in place. Before I install the sheet, I put a mark on the wall, or on an adjacent sheet, to remind me where that box is. When I tack this board in place, I make sure to keep the fasteners at least 2 ft. away from the box. Then I remove the lift, cut out for the box, and finish fastening the sheet.

Unless the room's width is a perfect multiple of 4 ft., the last row of drywall has to be ripped to size. This problem is not a concern when I'm working with a partner, but it's something I have to think about when I'm going solo. Because it's designed to handle large panels, the drywall lift doesn't work well if the sheets are narrower than 21 in.—or shorter than 50 in. I can avoid this problem by ripping a foot or so off the first row of drywall, or I can cut the last row into shorter lengths that I can manage by hand.

After all the full-length sheets are up, I go back and fill in the gaps. Boards that are too small for the lift are installed one of two ways, depending on size. For small sheets, I climb up on the bench, support the board with one hand, and drive a few screws using the screw gun in my other hand (photo, below). When I have to install larger sheets that are too heavy to support with one hand, I use my head (photo, above). (That's where a soft hat can be handy.) I adjust one

The author supports heavier sheets with his head and fastens them with a hammer to extend his reach.

of the step-up benches so that the top of my head just touches the joists while I'm standing on the top step. After I've wrestled the board into position, I use my head and my left hand to clamp it while I spin around and drive a few nails home using a hammer in my right hand; in this instance, I use nails rather than screws because the hammer enables me to extend my reach.

A Drywall Lift Is Not Just for Ceilings

When I turn to the walls, I have the option of placing the sheets vertically (standing them up) or horizontally (laying them down). Standing sheets up is easy to do (sidebar, p. 110), but I generally prefer to lay them down. For residential projects, rooms tend to be small, so laying the sheets down usually results in fewer seams to tape.

Wherever a wall is longer than 12 ft., laying the sheet down means I have to deal with a butt joint. Butt joints are a lot harder for tapers to make disappear, so I try to minimize those joints by making them fall above doors and above or below windows. When I don't have a door or window to help me, I stagger the seams as far from the center of the wall as possible.

Lightweight sheets are supported with one hand and fastened with a screw gun.

TIP

Wherever I encounter a recessed light or an electrical box, I avoid the hassle of measuring and fitting by using a router to make the cut after the sheet is tacked in place.

Standing Sheets Up

In most cases, I prefer to hang the wall panels horizontally, but there are situations where vertical is better. For instance, on a wall less than 4 ft. wide, I can eliminate a horizontal joint by standing a full sheet straight up. Standing sheets up also enables me to eliminate troublesome butt seams. If I've got a long, highly visible wall that must finish perfectly flat, I stand the sheets vertically and leave the finisher to work with a bunch of recessed edges rather than protruding butt seams.

For ease of placement, I cut each sheet at least ½ in. shorter than the height of the wall. I use the "tilt and lift" procedure (described in the main text) to pick up and carry the sheet where it needs to go. When I reach my destination, I set down the corner about 6 in. in front of the wall, stand the sheet, and leave the top edge of the sheet temporarily leaning against the wall.

After making sure that my foot pedal is within arm's reach, I grasp the sheet around waist height, lift it slightly off the floor, and then push the bottom of the sheet against the bottom plate of the wall (left photo).

Holding the sheet against the wall with one hand, I reach for the foot pedal with the other hand, place the pedal under the center of the sheet and apply foot pressure until the top of the sheet meets the ceiling (right photo.).

Gentle pressure on the pedal steadies the sheet while I center the tapered edge on the stud. I then drive a few screws to hold the sheet in place, remove the foot pedal, step back, and drive the remaining screws.

Tilt backward and lift from waist height to position a large vertical sheet.

Steady the sheet with one hand until it's held tight to the ceiling with foot pressure.

I hang the uppermost row of boards first. After cutting the first sheet to length, I get on the dark side of the board, set it in the clips, and roll the lift alongside the wall. After setting the floor brake, I tilt the top of the sheet away from the lift and let it rest against the wall. Then I crank the wheel until the top edge of the sheet kisses the ceiling. After fine-tuning the position, I crank the sheet tight (photo, below) and drive a handful of screws along the top edge to tack it in place. I then remove the lift and drive the rest of the screws.

I hang all the upper sheets in the room before I start on the lowers. I work my way around the room in a circular pattern, taking full advantage of the available tolerances. Whenever I reach a corner, I cut the board at least ¼ in. short, knowing that the adjoining board will cover the gap. The only sheet that must be cut tight is the last one.

When I have to hang a board that's too small for the lift (less than 50 in. long), I start a couple of nails in the top before I pick up the board. After it's in place, I steady the board with one hand and hammer

Sources

The Stanley Works
Goldblatt Tools
800-262-2161
www.stanleyworks.com
Foot pedal

Telpro Inc.
7251 S. 42nd St.
Grand Forks, ND
58201
800-448-0822
www.telprodirect.com

To hang the top panel on the wall, the author wheels the drywall lift up against the wall and sets the foot brake; while gently holding the sheet against the studs with one hand, he uses the other hand to crank the wheel until the top edge just touches the ceiling. After checking to make sure it's positioned correctly, he cranks the sheet tight.

home the nails with the other. Compared with screwdriving, the extra reach I get using hammer and nails allows me to tack the board without leaving the floor.

Foot Pedal Lifts the Bottom Sheet

Once the top row of sheets is installed, I'm finished with the drywall lift. I use a low-tech foot pedal to position the bottom row (top photo). After the first sheet is cut to

Raising the bottom layer. Because a standard drywall lift is too high off the ground to hang the bottom panel, a specially designed drywall foot pedal provides the leverage to close the gap between sheets.

length, I check to see if there are any electrical outlets on the wall. If so, I put a mark on the floor directly beneath the outlet and measure the distance from the floor to the center of the box (electricians always place their boxes at a uniform height, so I jot down this measurement to use for future reference).

If I have to move a sheet a long distance, I use the "tilt and lift" method, carry it where it needs to go, then reverse the lifting procedure to set the sheet gently on the floor. If the distance is only a few yards, such as across the room, I use what I call the "armpit carry": I center myself on the sheet, bend my knees, and tuck the edge under my armpit. I grip the sheet tightly using both hands as well as squeezing under my arm; then, keeping my back straight, I use my legs to lift the sheet off the ground. I could make things even easier on my back if I used a dolly to move the panels around (sidebar, facing page). But a dolly would slow me down, and in my world, production is always an issue.

When I've gotten the sheet roughly in position, I slide the foot pedal under the center, and I raise the sheet until its tapered edge meets the one above it (bottom photo). I drive a handful of screws along the top edge to tack the board in place; then I kick the foot pedal out of the way and mark the location of any electrical boxes. I rout out the boxes before I finish fastening the sheet.

As a rule, I prefer to hang all the drywall in one room before I move to the next room. When the clock is ticking on a rented drywall lift, however, I often first put up all the sheets that require the lift. Then I'll drive like a crazy man to get the lift back to the rental shop before closing time. On the way back, I stop at a nice restaurant and treat the entire crew to dinner.

Pat Carrasco hangs drywall and is a freelance writer in Bozeman, Montana.

New Tools Cut Out the Heavy Lifting

A drywall lift makes it easy for one person to install 12-ft. panels of drywall, but getting the material from the pile to the lift is hard work. Fortunately, Telpro Inc., the company that makes the PanelLift, also makes a couple of nifty devices to ease loading and transporting.

The PanelLift Loader is a simple attachment that mounts easily on any PanelLift drywall lift (photo, above). Lifting a corner of the sheet 2 in. off the ground is all it takes to load the loader; from there, turning a crank handle raises the sheet and sets it gently in the support hooks.

The Troll is simply an inexpensive ($30) dual-wheel roller with a handle (photo, right). I've seen other types of panel-moving dollies, but this design also makes it possible to carry a panel over obstacles or to raise it high enough to be placed onto the PanelLift Loader.

A Dramatic Family-Room Addition

■ BY B. ALAN LEONARD

New Jersey, the Garden State. Every once in a while, you come across an area of New Jersey that actually lives up to the state's nickname. Jeff and Chris Wells recently moved to such an area. Set graciously on a gently rolling, 3-acre wooded parcel near a large parkland, their home is a 35-year-old, 1½-story French-style ranch that included details both graceful and shabby. The house united the fine detailing of arched copper dormers and chimney pots with rotting board-and-batten siding; elegant chimney pots were combined with corrugated roofing on a detached garage.

The Wellses had many wishes for their new home, and they asked me to design a renovation that included adding bathrooms and changing the siding. Their greatest desire, however, was for a new room that would take full advantage of their beautiful land as well as lead to a new patio where they could entertain outdoors. The new room need not be large, but Jeff and Chris wanted it to combine a sense of excitement and elegance that defined their lifestyle.

The existing den was the best choice for a renovation because it felt dark and was not open enough to the backyard. Our first thought was to remove the rear wall and enlarge this room. But the den's 8-ft. ceiling was a problem. Had we continued the ceiling at that height, no drama. If we raised the addition's ceiling, the resulting room would be too disjointed. Our solution was to build a separate family room (19 ft. by 19 ft.) adjoining the existing den.

Careful Alignment of Windows and Doors

The entrance to the den from the hallway is at one corner of the room, so the entrance to the new room aligns with the den entrance, making one side of the den into a natural circulation corridor. The family room itself is centered on its entrance, with the opposite wall containing a large window grouping that focuses on a swimming pool. Enclosed by a split-rail fence and elaborate plantings, the pool area exudes rural charm.

This visual axis is crossed by a second axis created by a stone fireplace on the north wall and a grouping of four French

This family-room addition offers
dramatic views from within and without.

No structural members pierce the cathedral ceiling. Thanks to a band of steel angle iron around the walls, the 18-ft.-high cathedral ceiling in this room addition remains unobstructed by collar ties. Arched dormers and white paint also contribute to the room's light and airy feeling.

doors on the south wall. Decorative oval windows were placed on both sides of the doors, which open onto a new flagstone patio (photo, p. 115).

The fireplace creates a strong termination to the view from the patio. Because stone is massive and heavy, the fireplace also helps to weigh down the height of the room.

Hipped Cathedral with Dormers Intensifies Structural Issues

The ceiling is high, nearly 20 ft. With arched dormers and no collar ties in the ceiling, the room is dramatic. Designing and building this roof offered challenges that appealed to me and to the framers, Kevin and Paul Delaney of Long Valley, New Jersey.

One of the strongest visual features of the main house is its mansard roof. Beginning at the second floor, the roof rises approximately 12 ft. at a 21-in-12 pitch, terminating in a flat roof. To maintain the character of the house, I designed an identical mansard roof for the addition. The new roof extends approximately 20 ft. above grade and is capped off with what is essentially a flat roof (photo, below). Had we carried the hip to its natural peak, it would have towered over the existing house, soaring out of proportion and violating the scale of the new room.

Inside, the roof shell is exposed, resulting in a hipped cathedral ceiling. I have designed many gable-style cathedrals, and they are relatively simple: Use a structural ridge beam supported at either end, or use some sort of exposed collar ties near plate height to keep the weight of the room from spreading the outside walls. A hipped cathedral, however, needs to resist splaying in two directions, so a structural ridge cannot be used. Exposed collar ties would work; however, they would need to run in two directions. Much too bulky for a room of this size, a gridwork of exposed ties would tend to feel like a ceiling. The roof had to be built without collar ties.

Bringing in natural light was also an issue. Most cathedral ceilings tend to feel dark and heavy unless light is introduced. In this roof, dormers not only illuminate the ceiling space, but they also add architectural planes that catch light differently throughout the day, creating a dynamic spatial quality in the room.

Outside, the dormers have a positive impact by breaking up the roof area. Their arched roofs and windows continue the design language of the existing house dormers. But the rough openings for the dormers would also concentrate the roof loads at certain points along the top plate and further complicate the framing.

A mansard roof is a hip with a flat top. Framed with 2x10 rafters that rise at a pitch of 21-in-12, this roof is capped by a square framework of doubled 2x10s. The steel angle that resists the rafters' outward thrust is visible along the top of the walls.

A Hip Roof with a Flat Top

Two factors complicated the framing of this roof. First, neither a structural ridge nor collar ties could be used to resist the outward thrust of the rafters. Instead, the walls are reinforced with 3-in. by 5-in. steel angle, sandwiched in the top plates, welded at the corners, and acting as a tension ring. Second, four dormers pierce the roof and concentrate point loads at certain points along the top plate. The solution here was to support the roof with two pairs of primary frames, or rafters, linked by 4x8 roof joists.

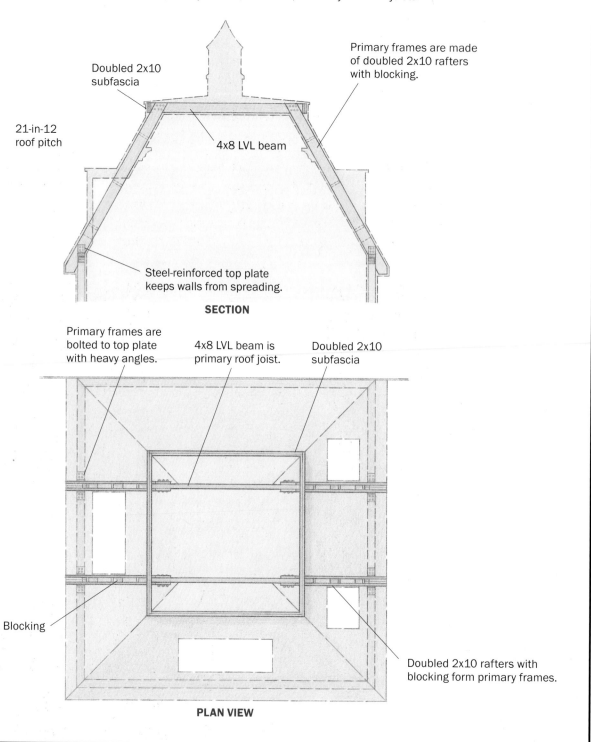

Doubled 2x10 subfascia

Primary frames are made of doubled 2x10 rafters with blocking.

21-in-12 roof pitch

4x8 LVL beam

Steel-reinforced top plate keeps walls from spreading.

SECTION

Primary frames are bolted to top plate with heavy angles.

4x8 LVL beam is primary roof joist.

Doubled 2x10 subfascia

Blocking

Doubled 2x10 rafters with blocking form primary frames.

North ▶

PLAN VIEW

Drawing by Vince Babak

A Steel Tension Ring Keeps the Walls from Spreading

The purpose of collar ties is to resist the outward thrust of the rafters and to prevent the exterior walls from splaying out. Eliminating collar ties meant we had to come up with another way of keeping the walls plumb. A steel tension ring was the answer. Sandwiched into the walls' top plates, a band of steel angle iron reinforces the walls, and it prevents them from splaying.

The Delaneys framed the 9-ft.-high walls with 2x6 studs, which are sturdier than 2x4s and allowed for additional insulation. The 2x6s for the double top plate were ripped down ½ in. to provide clearance for the continuous 3-in. by 5-in. by ½-in. steel angles.

The Delaneys' crew lifted the angles into place and welded them together at the corners (photo, below). A third 2x6 was added to provide nailing for the rafters; then the entire plate was through bolted. I will always remember seeing Raphael, one of the carpenters, sitting on his drill atop the wall, boring holes through the steel-reinforced top plate.

Core of Roof Frame Is a Pair of Oversize Rafters and Joists

Working from scaffolding set up in the new room, the crew framed the roof starting with a pair of oversize rafters and joists, or primary frames, that flank the dormers on the north and south slopes (drawing,

Welded corners keep the walls plumb. The top plate contains continuous 3-in. by 5-in. by ½-in. steel angle between layers of 2x6s. The entire plate is through bolted, and the corners are butted together and welded.

on p. 118). Each primary frame consists of two 2x10 rafters bolted to a 4x8 laminated veneer lumber (LVL) beam serving as a roof joist. The primary frames, spaced with blocking along their length, are bolted to the plate with heavy angles (photo, below). Then the primary frames were tied together with a subfascia of doubled 2x10s. Other than between the primary frames, all roof framing is supported by these two frames.

Due to the location of the dormers flanking the chimney on the north wall, we could not align the primary frames with the sides of the flat roof. Instead, the frames are set back into the flat section, and a system of 2x8 outriggers extends out from the 4x8s to the subfascia. The outriggers pick up the 2x10 rafters on the east and west slopes, which are connected to the outriggers with clinched 20d nails. Clinching means using oversize nails that punch through the wood

and are bent to keep all layers tight. There are double 2x8 outriggers at the corners that pick up the hip rafters (photo, facing page). All rafters are joined to the wall plate with framing anchors, as are the outriggers to the 4x8s and the subfascia.

The flat roof is framed with 2x8s running parallel to the primary frames with solid blocking between to stiffen the frame. Strips of 2x material ripped on an angle provide for drainage. (The flat roof's pitch is actually about 1-in-12.) These strips also create a space above the roof insulation for air to flow from soffit vents to a cupola. Not only does the copper-roofed cupola handle the venting, but it is also a visual tie-in to the existing cupola on the garage roof.

By stiffening the connections between the roof frame and the wall as well as between the flat roof and the sloped rafters, we have a strong system that resists the out-

Dormers create concentrated loads. The main structural supports for this roof are the doubled rafters, or primary frames, that flank the dormers. At the peak, these rafters sandwich a 4x8 roof joist, which is why they are spaced apart by blocking. Simpson™ HL76 steel angles tie the primary frames to the wall plates.

Hip rafters tuck under the subfascia. A platform of crisscrossing 2x8s, ringed by a double 2x10 subfascia, caps off the mansard roof. At the top, the doubled-up hip rafter is sandwiched and carried by a pair of 2x8 outriggers running on a diagonal.

ward thrusts of the roof. The steep pitch also helped; snow loading is limited to the flat section. The sloped sections are subject more to wind loads than to snow loads.

Bands of Trim Mediate Ceiling Height

Although the roof dormers bring light into the upper reaches of the ceiling, more needed to be done to temper the height of the room. An 18-ft.-ceiling in a 19-ft. by 19-ft. room feels high! Two oversize bands of moldings—one 12-in.-wide crown detail at plate height and another 7½-in.-wide crown at approximately 15 ft. above the floor— break the ceiling into smaller sections.

Constructed of stock components, both crowns are blocked out from the walls and roof to allow for concealed lighting. I specified continuous incandescent strip lighting on separate dimmer circuits. The top band

lights the flat ceiling, and the lower band lights the sloped ceilings.

Installing a 6-in. colonial base and a 4-in. chair rail at 36 in. above the floor gives added visual weight. The stone-veneer fireplace with its large wood mantel anchors the room as does the dark green paint on the walls with a lighter ceiling. As a finishing touch, there's a walnut strip border in the oak flooring.

The new room is comfortable for entertaining. People tend to gravitate to it both for the views outside and for the views inside. The most satisfying comments suggest that the room doesn't just fit the house, it completes it. Such is the goal and the reward.

B. Alan Leonard is an architect in New Providence, New Jersey.

A Well-Lit Addition

■ BY ROBERT L. MARX

Often, the challenge of renovation is finding the solution that provides for the clients' new needs while preserving what they enjoy in their home. This project introduced an ironic twist, for the home's most distinctive elements also contributed to its shortcoming.

The Dilemma

My clients loved their home, especially for its ivy-covered fieldstone walls and its arched windows. An old estate carriage house in Greenwich, Connecticut, the building had been carefully converted into a residence. Although the layout of the first floor made the best of three arched windows on the west elevation, the rooms on the first floor were dark. My goal was to expand the house and to brighten the interior. However, the plan made expansion impossible without disturbing the fieldstone walls, and the owners wanted to leave both the walls and the arched windows intact.

So I decided to add living space at the back of the house (photo, p. 126). I connected the addition to the house by removing an earlier addition that contained two small, unused rooms. Doing away with the old addition increased the area available for new construction and provided a direct connection between the new family room and the rest of the house. More importantly, it preserved the old stone walls.

However, the addition's northern exposure does not provide the most advantageous solar orientation, so I had to search for ways to capture natural light.

Reaching for the Light

Because I was using a combination of glass doors, transoms, and overhead glazing to light the addition, my goal was to arrange the glazing into a composition in keeping with the house.

The windows and the doors are similar to those of the original house not only stylistically but also in how they were installed. On the interior of the original house, the thickness of the fieldstone walls is apparent because windows and doors are set deep in their openings. I wanted the addition to have the same character, even though the new walls would be framed with 2x6s. To simulate the thickness of a stone wall in the addition, the triple window and door units project 1½ ft. beyond adjacent wall planes, creating a bay window and a door alcove.

A glazed cupola and exterior transoms let daylight into a new family room with northern exposure.

Both of these projections have flat roofs. Each flat roof is lined with a lead-coated copper pan drained with scuppers.

Although 1½-ft.-deep window and door openings help unify the original house and the addition, the bay window and the door alcove reduce the amount of daylight coming into the room. To overcome this loss of light, transoms were installed in the same plane as the framed walls. The window transom is framed in the gable-end wall. The door transom forms a shed dormer rising from the main roof.

Even with the transoms, the lighting was inadequate, so I concentrated next on the roof. The owners and I thought that skylights would detract from the character of the old carriage house. We all agreed that a

Section of Roof Framing

Within the gable roof, two headers support the cupola; these headers are fastened to doubled rafters at each end of the cupola. A tripled header carries the cripple rafters at the shed-dormer opening. This header is fastened with bolts and clip angles to tripled rafters on each side of the door alcove. At the top of the alcove opening, the lower transom header hangs from a pair of wood 2xs that are bolted to the tripled rafters; this header is concealed in the soffit inside the room.

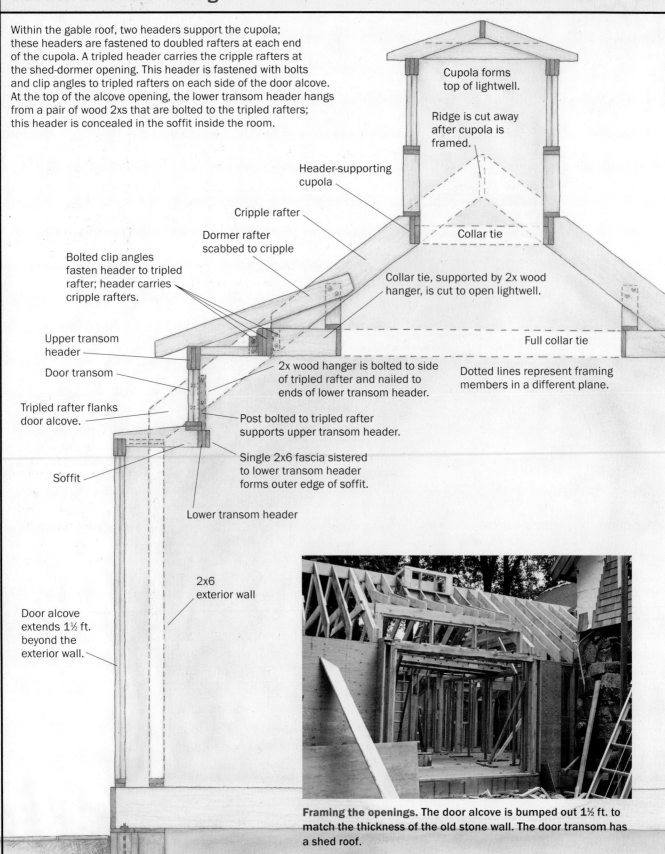

Cupola forms top of lightwell.

Ridge is cut away after cupola is framed.

Header-supporting cupola

Cripple rafter

Dormer rafter scabbed to cripple

Collar tie

Bolted clip angles fasten header to tripled rafter; header carries cripple rafters.

Collar tie, supported by 2x wood hanger, is cut to open lightwell.

Upper transom header

Full collar tie

Door transom

2x wood hanger is bolted to side of tripled rafter and nailed to ends of lower transom header.

Dotted lines represent framing members in a different plane.

Tripled rafter flanks door alcove.

Post bolted to tripled rafter supports upper transom header.

Single 2x6 fascia sistered to lower transom header forms outer edge of soffit.

Soffit

Lower transom header

2x6 exterior wall

Door alcove extends 1½ ft. beyond the exterior wall.

Framing the openings. The door alcove is bumped out 1½ ft. to match the thickness of the old stone wall. The door transom has a shed roof.

glazed cupola would contribute more to the character of the house. The cupola, placed at the center of the ridge, lends warmth to the new family room by transmitting diffuse overhead light (photo, p. 123).

The French Correction

Like the windows of the original house, the new ones are double-hung wood with true divided lites. The lite pattern of the French doors on the west elevation matches that of the windows.

I prefer the feel and the operation of French doors over sliding doors. However, manufacturers have standardized the width of French-door stiles at more than 4½ in. I planned to group three doors together and thought these stiles would look heavy. With narrow backset mortise locks easily available, other door-sash dimensions are possible.

We had narrow-profile doors with thick sections custom built by the Woodstone® Company. These 2-in.-thick doors have 3½-in.-wide stiles, which approaches the slenderness of sliding-door frames and is particularly appealing when several doors are used in combination. Like the windows, the doors are single glazed with true divided lites. With their storm panels in place, the doors provide an R-value equal to doors with ½-in. double glazing. Only the center door is operable; the others are fixed.

I sometimes use laminated glass, as I did here, rather than tempered glass where safety glazing is required by code. In larger sheets, laminated glass doesn't reflect wavy images as tempered glass sometimes does.

Headers Abound in the Roof

Like the roof of the old house, the new roof is a gable. But its framing was complicated by both the cupola and the shed dormer for the door transom. Tripled rafters on both sides of the door alcove carry the header for the shed dormer (photo and drawing, facing page), which is fastened with clip angles and bolts. Doubled rafters support headers that carry the cupola. Once the cupola was framed, a 42-in. section of the ridge beam was cut between the doubled rafters to open the lightwell below the cupola.

The remaining rafters are single 2x12s. Their collar ties create a flat ceiling plane 10 ft. above the floor. In the center of the ceiling, there's a 7-ft. opening for the light-well. The walls of this opening are flared, with the east and west walls of the opening formed by the underside of the rafters. The north and south walls of this well are supported by headers framed on the collar ties of the tripled rafters. These walls are sloped to match the angle of the rafters, creating a symmetrical, flared lightwell.

A soffit runs along the east and west walls. The plane of the bottom of the rafters is visible above this soffit. The soffit provides a smooth transition from the door transom to the roof framing and gives an intimate scale to the room.

Trim Inside and Out

I unified the various elements of the room—the soffits, the lightwell, the window seat and the door alcove—with a picture rail. This trim runs continuously around the room, along the bottom edge of the soffit and between the transoms and the doors and the windows, forming the top casing of the doors and the windows.

The 2x12 rafters provide ample room for R-30 batt insulation and a 2-in. airspace to ventilate the underside of the roof deck. But neither the depth of the framing nor its spacing match that of the original house. The house reveals its 2x6 rafter tails and T&G beadboard roof decking at the eaves behind a 2x6 fascia board. The addition is also detailed with 2x6 rafter tails placed to match the spacing of those on the house. Within the roof framing, the inboard ends of the 2x6s are nailed into blocking that spans between the rafters. The 2x6s bear on

Sources

Baldwin® Hardware
841 East
Wyomissing Blvd.
Reading, PA 19612
800-566-1986
www.baldwinhardware.com
Backset mortise locks

Marvin Windows and Doors
P.O. Box 100
Warroad, MN 56763
800-537-7828
www.marvin.com

Woodstone Company
P. O. Box 223
Westminster, VT 05158
802-722-9217
www.woodstone.com

TIP

Try using laminated glass instead of tempered glass where safety glazing is required by code. In larger sheets, laminated glass doesn't reflect wavy images as tempered glass sometimes does.

Solving two problems. Locating the addition on the north end of the house left the fieldstone walls of the original house intact. Combining triple French doors, triple window units—both of which are topped by transoms—and a glazed cupola tastefully illuminates the new room.

another row of blocking along the top plate, resulting in cantilevered dummy rafter tails on the addition that match the exposed rafter tails on the main house.

Let the Stone Show

Details sometimes go unnoticed when they work well. One example is where the addition roof butts into the stone wall of the original house. The easiest solution to this detail—in fact, the one that had been used on another part of the house—would have been to extend the shingle siding down from the second floor over the stone on furring strips until it met the lower roofline. The sidewall flashing could then be turned up behind these shingles. But we didn't want to cover up the stone wall and didn't have to, thanks to the skill of builder Wayne Dudley of Weston, Connecticut. Dudley cut

a continuous groove, or reglet, directly into the stone to flash this roof-to-wall intersection. He used an abrasive sawblade to cut the 1½-in.-deep groove. Lead-coated copper flashing is held in the groove with shims and is finished with a bead of sealant. This subtle, clean detail allowed us to expose the fieldstone wall to the east and as it turns the corner.

Robert L. Marx is an architect in Stratford, Connecticut.

Converting a Garage into Living Space

■ BY NEIL HARTZLER

When my wife and I postponed the building of a custom home for ourselves, it was on condition that we would instead add two major features into our present residence: a library/study and a large two-car garage with workspace for my tools. Located in Colorado Springs, Colorado, our neighborhood does not provide the outdoor space we wished for in a new house, but it has been a pleasant place to live.

By working weekends and evenings we built the new freestanding garage in the back of the house, allowing access to an adjacent alley. Then, over a period of five months, my wife and I turned the old attached, single-car garage (top photo, p. 128) into a library and study (floor plan, p. 130). Being a cabinetmaker, I looked forward to making all the trim and casework for the study. But first we had to solve three problems typical of a garage conversion: (1) How do you keep the facade from looking like a converted garage? (2) What do you do about the built-up beam that supports the ceiling joists? (3) How do you deal with the slab floor?

First, a Face-Lift

We were very concerned with the exterior appearance of the house following the conversion. Many times I've seen a garage conversion accomplished by the installation of a sliding patio door in the space formerly occupied by an overhead garage door. This approach reinforces the converted-garage appearance, especially if the old driveway is not removed. To avoid this problem, we decided to build a floor-to-ceiling bay where the garage doors had been and to continue the existing brick facade around the bay capped by a hip roof (bottom photo, p. 128).

Leaving eave design and soffit and header heights consistent with existing roof and windows provided continuity, yet the angled bay walls helped break up the straight facade. The angle bay also provided the additional floor space we needed for a desk and a chair. Once we took out the driveway and landscaped the space where it had been, all evidence of the original garage was eliminated.

Ready for a remake. After building a freestanding garage in the back of his property, the author was ready to convert his one-car garage (shown here) to a library/study.

Blending in a bay. A full-height bay window replaces the overhead garage door and adds visual interest to the simple brick facade. To avoid the garage conversion look, the author dug up the old driveway and beefed up the landscaping.

The New Ceiling Structure

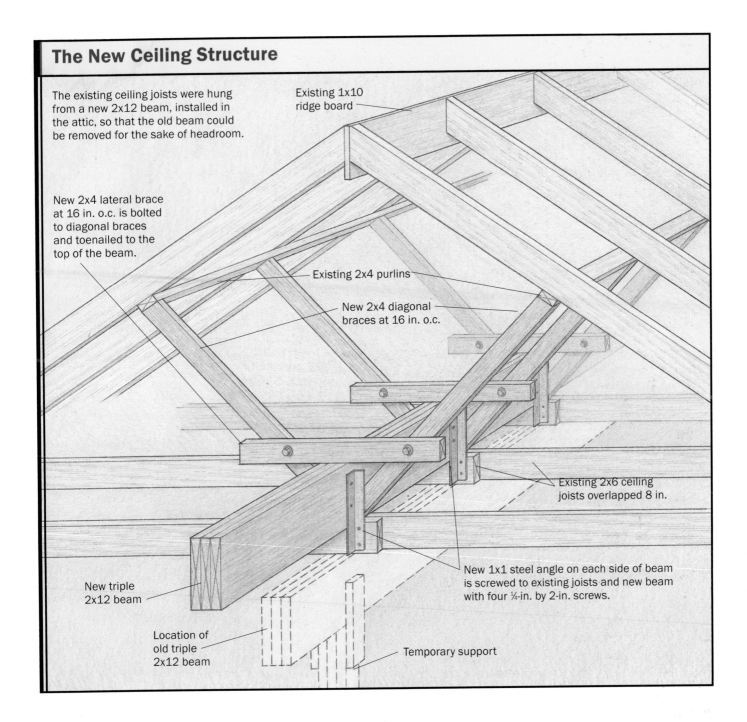

The existing ceiling joists were hung from a new 2x12 beam, installed in the attic, so that the old beam could be removed for the sake of headroom.

Existing 1x10 ridge board

New 2x4 lateral brace at 16 in. o.c. is bolted to diagonal braces and toenailed to the top of the beam.

Existing 2x4 purlins

New 2x4 diagonal braces at 16 in. o.c.

Existing 2x6 ceiling joists overlapped 8 in.

New triple 2x12 beam

New 1x1 steel angle on each side of beam is screwed to existing joists and new beam with four ¼-in. by 2-in. screws.

Location of old triple 2x12 beam

Temporary support

Second, a Structural Overhaul

The ceiling of the existing garage was composed of 2x6 joists that were lapped over a triple 2x12 beam exposed below the ceiling line. The existing beam would be too low once we framed a new floor over the old garage floor, and it would interfere with the space we would need for a header over the new French doors into the house. To eliminate the beam we had to find another method of supporting the joists.

We solved this problem by building another beam above the joists and hanging the joists from it (drawing, above). This beam had to be built in place because there was no way to get a completed beam into its space without cutting a hole in the end of the house; we considered this method only as a last resort.

The ends of the original beam rested on cripple studs, directly above where we wanted the French doors and the fireplace to go. We would not be able to install the new beam until the headers for these openings were in place. And we couldn't install the new headers with the old beam in our way.

To get around this problem, we decided to support the old beam temporarily on post so that we could cut off the ends and insert the new headers. We placed tripled 2x4s under the existing beam and braced them securely to the existing frame. We prevented the bottom of each column from moving by bracing it to the existing bottom plates of the garage walls. Then the ends of the old tripled 2x12 beam were cut off and removed. This allowed for new framing over the new doorway and fireplace openings.

In the existing wall between the garage and the house, there wasn't enough maneuvering room to install a conventional header built with 2x8s and a ½-in. spacer. Instead, we built a header out of plywood, in effect making our own laminated beam in place. We glued five layers of fir plywood together and, after the glue had set and the clamps

A New Fireplace and Bay Window

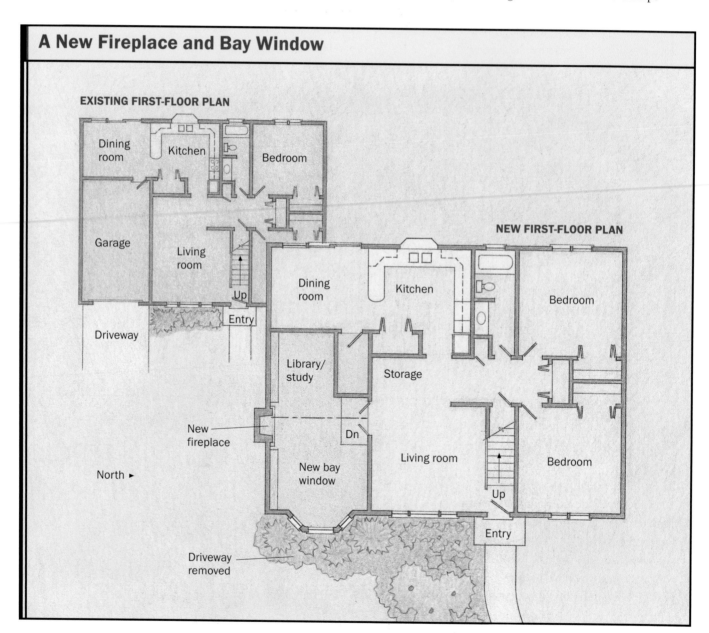

EXISTING FIRST-FLOOR PLAN

Dining room

Kitchen

Bedroom

Garage

Living room

Up

Entry

Driveway

NEW FIRST-FLOOR PLAN

Dining room

Kitchen

Bedroom

Library/ study

Storage

New fireplace

Dn

Living room

Bedroom

North ▶

New bay window

Up

Entry

Driveway removed

were removed, nailed the laminated beam from both sides with three rows of 16d nails, 8-in. o.c. I've found the best glue for such applications is a waterproof product manufactured by National Casein® called Cross-Link Adhesive, which employs both a liquid catalyst and a liquid resin.

The door header spanned 52 in. The header over the fireplace, which we built the same way, spanned 44 in. We installed the door header tightly against the underside of the top plates of the existing wall. We dropped the fireplace header to rest just above the zero-clearance firebox. Then we installed cripple studs between the plywood header and the existing top plate, with triple studs used directly beneath the bearing point of the new beam.

Using three new 2x12s, we assembled the new beam in the attic over the new room and installed it over the existing top plates, directly over the location of the old beam. We used angle steel and lag screws to fasten the old joists to the new beam (drawing, p. 129); we toenailed new diagonal braces to the roof purlins and the new beam. To prevent further twisting or movement of the beam, we bolted horizontal 2x4 struts to the 2x4 diagonal braces (perpendicular to the new beam) and toenailed the braces to the top of the new beam. Finally, we removed the old beam and all temporary bracing.

An Insulated Floor over the Slab

Before dealing with the floor over the slab, we built a 4-ft. by 7-ft. closet in the corner of the garage for a freezer and gave the space a door to the existing dining room. We insulated the closet walls, as well as the existing wall between the dining room and the new library/study.

The garage slab sloped toward the driveway, and we could have built a level floor by putting down tapered sleepers with plywood over them. But we wanted plenty of insulation underfoot, so we suspended a conventional floor frame, with 2x6 joists, over the slab. We lag screwed 2x6 rim joists to the existing walls around the perimeter of the room. The difference in floor level between the living room and the garage was 12 in., so the top of the rim joists were still well below the level of the existing living-room floor. A 6-in. landing at the French doors made up the difference, allowing a greater ceiling height in the library.

We framed the floor with 2x6 joists hung on joist hangers that were fastened to the rim joists. The total span was 12 ft., so we wedged metal shims between the floor joists and the existing concrete floor in the center of the span to prevent deflection. These 2-in.-wide by 3-in.-long pieces of metal—we call then framing shims around here—are used wherever there's a gap between framing members. Our inspectors routinely call for them, particularly where bearing surfaces do not meet exactly.

After floor joists were in place, we ran electrical feeds in the floor and the walls. We ran a gas line from the crawlspace under the main house and later hooked it up to a gas-log unit in the fireplace.

I considered placing a vapor barrier over the concrete but decided against it because we never had moisture problems in the garage. Besides, the air here in Colorado is pretty dry. If you live in a more-humid part of the country, a vapor barrier would be a good idea. We stapled 6-in. fiberglass batts between the joists, then installed the subfloor. We installed ¾ T&G waferboard subflooring using 8d ring-shank nails and panel adhesive.

Textured Walls

Before hanging drywall, we added 3½-in. fiberglass batt insulation to the new walls, including interior walls, which would soundproof our study somewhat. We had to use ⅝-in. drywall to patch the ceiling where we had removed the old beam. And we had to patch existing walls to match the existing

Sources

Fuller O'Brien Paints
450 E. Grand Ave.
S. San Francisco, CA
94080
415-761-2300
www.fuller-obrien.com

The Majestic Co.
1000 E. Market St.
Huntington, IN 46750
219-356-8000
www.
majesticfireplaces.com

National Casein Co.
601 W. 80th St.
Chicago, IL 60620
773-846-7300
www.
nationalcasein.com

Robert H. Peterson Co.
14724 E. Proctor Ave.
City of Industry, CA
91746
800- 332-3973
www.rhpeterson.com
Real-Fyre Gas Log Set

⅝-in. fire-rated drywall. On new walls, we used ½-in. drywall.

Once the drywall was hung, taped, and sanded, we textured the ceiling and all wall areas that would be exposed and painted. Areas to be papered were sanded as smoothly as possible. We used two different textures in the library; both were applied with a damp sponge, using a slightly thinned mixture of joint compound. To texture the ceiling we dipped the sponge in the compound mixture, then applied it to the ceiling in a sweeping motion. This technique left a series of thin raised lines that look much like heavy, random brush marks.

When we applied the texture to the walls, we again used a sponge, but this time the thinned compound was daubed on the wall by pressing the sponge to the surface of the wall, then pulling it away. To add variety we occasionally gave the sponge a twist, then pulled it away.

Adding On a Fireplace

Our home had no fireplace prior to this addition. Although a fireplace is a feature much sought after in the Colorado Rockies, there is also a great deal of concern about air pollution here. With the possibility of restrictions on wood-burning fireplaces and stoves on the horizon, we chose to install a gas-log fireplace.

For the fireplace box, we chose the Majestic MBUC 36 brick-lined wood-burning unit. To accommodate burning wood, we installed a triple-wall flue system. We had a sheet-metal cover made to fit the top of the flue chase and then used a standard storm collar and flue cap on the flue end.

We purchased the gas-log unit for $240 from a local dealer. The unit includes a set of ceramic logs, a fire grate, a gas burner, a valve and pilot unit, and a length of brass fuel line to be attached to lead pipe fittings. Installation of these units is not cheap— $300 in our case. At this point, the cost of this fireplace, without any of the framing or finish materials included, was $1,300. The fireplace was a costly part of the project, but we're glad we did it. We enjoy the fireplace a great deal, and it has certainly added to the value of our home.

Cherry Casework

I chose cherry for all the casework and trim in this room (photo, facing page). Cherry is my favorite domestic wood because of its color and figure, but it is not easy to work with. Cherry is of medium hardness, the grain can be very difficult to read, and it tears easily if planed in the wrong direction. This tendency to tear is aggravated by humidity changes, even if the cherry is properly dried to a point of stability. Cherry also is a relatively expensive domestic hardwood. If I had built the casework out of oak, my lumber cost would have been about 60% of what I spent on the cherry.

I built and installed a cherry fireplace surround, with the tile recessed 2 in. from its face. To make the mantel, I stacked shaped pieces of cherry to create the effect of a crown molding (drawing, facing page) and returned each piece to the wall. The flat top of the mantel was attached with glue and concealed screws to the top of this built-up molding. I made all these parts with a table saw and a router. I've found that many interesting and diverse moldings can be created by building up different profiles made with just these two tools.

I constructed the bookcases with 4-in.-wide fluted stiles. Where necessary, I made double case sides so that the stiles overlapped the case sides by only ⅛ in.; this way no books are hidden behind stiles. The adjustable shelves are supported by metal shelf standards mortised into the case sides. The crown molding around the top of the bookcases matches that on the mantelpiece.

Prior to installation, I finished the casework with a cinnamon stain and a sprayed catalytic finish called Fullerplast from Fuller O'Brien® Paints. It depends on a chemical reaction for curing and is more water-resistant than a conventional lacquer finish is.

Cozier than a car park. The author's site-made, built-up cherry casework completes the garage conversion. A gas-log fireplace adds heat and ambience without the possible pollution of a wood fire.

It's a durable finish, but drying and recoating times are longer, and it's fairly expensive. I could have cut my finishing costs considerably by using a conventional lacquer. In the future I'll restrict my use of the material to applications where there will be water or moisture, such as on wooden bar tops or bath cabinets. After doors, casework, and trim were in place, carpet was installed, completing the room.

*Cost estimates from 1992

Neil Hartzler *has been a custom cabinet and furniture maker for 30 years. He lives in Colorado Springs, Colorado.*

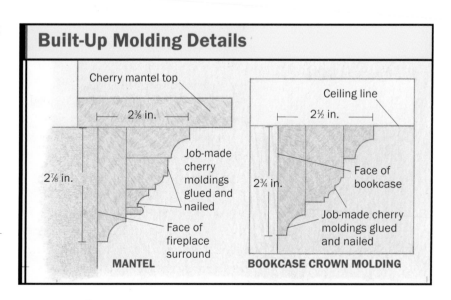

Built-Up Molding Details

Cherry mantel top

2⅜ in.

2⅞ in.

Job-made cherry moldings glued and nailed

Face of fireplace surround

MANTEL

Ceiling line

2½ in.

2¾ in.

Face of bookcase

Job-made cherry moldings glued and nailed

BOOKCASE CROWN MOLDING

A Suburban Metamorphosis

■ BY JOHN AND ELIZABETH INESON

Evidently, our home's builder had a great respect for several revival styles because he made sure that many were well represented. Built in the 1920s during a suburban cottage-revival period, the house combined gambrel rooflines and a brownstone base with the Tudor styling of fake timbers and stucco (top photo, facing page). The color scheme ranged from brown to browner. The fact that we are both architects made us more sensitive to the problems. After living in the house for six years, we decided we'd had enough.

Our plan was somehow to unify the exterior styles. We decided to replace the pseudo-Tudor trim with a consistent shingle-style interpretation and to alter the line of the gambrel roofs. At the same time, we planned both to enlarge the master bedroom and to add an office space in the attic.

A New Roof Over the Existing Roof

Our proposed revision of the existing gambrel roofs on both sides of the main house generated a good deal of discussion early on in the project. We were anxious to change the appearance of the roof, but were daunted

Modified rooflines help to unify the design. The original gambrel rooflines were straightened and the siding materials simplified to create a more appealing, practical design (bottom photo). The original house (top photo), built in the late 1920s, reflected the popular revivals of colonial, Dutch gambrel, English Tudor, and shingle styles. Unfortunately, this house displayed too many styles at once.

Reshaping the Rooflines

To hide the original gambrel roof, the authors wanted to extend the upper slope of the roof and to make a uniform 10-in-12 pitch. Because they were living in the house during the renovation, it made sense simply to build over the old roof. The detail at right illustrates the flashing used to protect the dormers, now recessed behind the roofline.

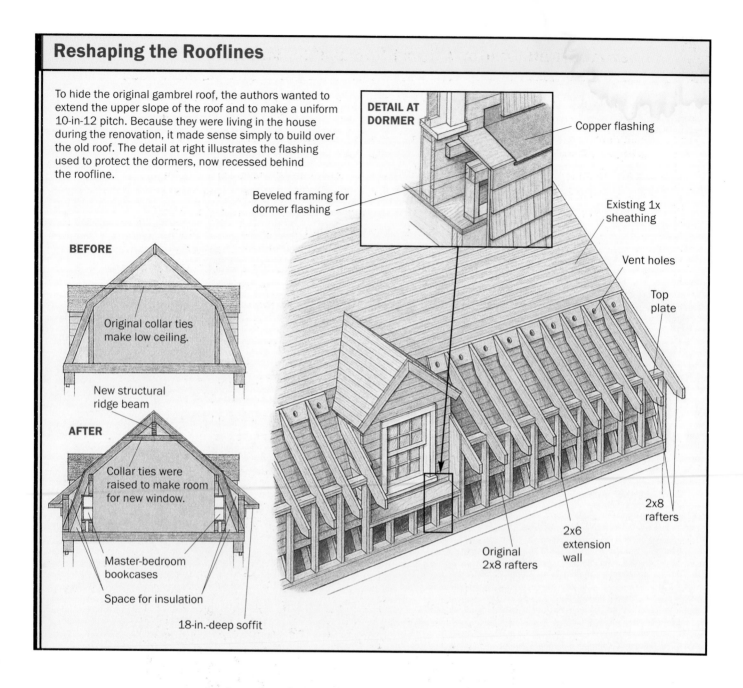

DETAIL AT DORMER

Copper flashing

Beveled framing for dormer flashing

Existing 1x sheathing

Vent holes

Top plate

BEFORE

Original collar ties make low ceiling.

New structural ridge beam

AFTER

Collar ties were raised to make room for new window.

Master-bedroom bookcases

Space for insulation

18-in.-deep soffit

Original 2x8 rafters

2x6 extension wall

2x8 rafters

by the potential cost and extent of the work. After some deliberating and sketching, we proposed extending the gambrel's upper roof down and raising the sidewalls up to meet the eave. The construction of this new roof was complicated by the fact that we were living in the house. We got around this potential problem by leaving the original roof intact and building the new structure over it.

Our contractor, Tom Nedweden, removed only enough roof sheathing to make room for the new 2x6 wall extension (drawing, above). After building the sidewall to its new height, the crew pitched the new 2x8 rafters to match the existing upper roof slope of 10-in-12. At the eaves, the rafters were extended past the top plates to create an 18-in.-deep soffit with a continuous vent. Holes drilled through the sheathing in each existing rafter bay allow proper airflow up to the ridge vent. New soffit and ridge vents also help outside air to circulate past the roof surface, reducing heat transfer from the interior to the eaves during the winter months. The new roof structures were also a good

Custom Storm Windows Sport Removable Screens

As part of the renovations for our house, we removed the existing triple-track aluminum storm windows in favor of a more historically appropriate replacement. We wanted to reproduce the look of the original painted wood storms, but we knew that separate sets of wood storm and screen sash would be too expensive. We also didn't want to spend weekends removing one set and hanging the other at every change of season.

Working with Marvin Doors and Windows, we designed a storm sash with a fixed, glazed upper lite and a removable lower lite (photo, below left). The lower opening can be filled with either an aluminum-framed glass panel or a screened panel held in place with thumb turns that make replacement a quick task (photo, below right). The sash are hung on the exterior window casing in the traditional manner with two storm-sash hangers at the head and a hook-and-eye on the inside face of the sill.

The cost of the windows was more than triple-track aluminum storms; here in the Northeast, aluminum storms typically cost between $50 and $80 per window, depending on their quality. The wooden storms averaged out to be about $135 each. Although the price may seem steep, we think that the advantages of the windows definitely justify the cost. The custom storms give the house a more historically consistent look, plus they provide a tighter seal at the windows, unlike some aluminum storms. We don't even mind having to store one set of panels while the other is in use; they certainly don't take up much space. Our big regret is that we had the sash primed only and not finish-painted as well.

Custom storm windows combine tradition with interchangeable lower panels. Lower storm panels can be easily replaced by screened panels with the twist of a thumb turn.

excuse to install a self-adhesive bituminous membrane, a good design practice in this climate because the membrane helps to eliminate water problems related to ice dams at eaves.

As soon as the roof was finished, we stripped off the old cedar shingles and other trim. Assisted by a strong windstorm, the contractor removed the existing building paper and installed Tyvek® on a calmer day. We used copper flashing at the heads of all windows, on top of horizontal cedar trim and for masonry flashing, and 14-in.-wide window pans for the recessed dormers.

Custom Stress-Skin Header Carries Microlam in a Tight Space

The Inesons' plan to enlarge the master bedroom in their house included raising the collar ties in the ceiling, adding a structural ridge beam, and installing a large arch-top window. When they realized that the traditional header needed to carry the ridge beam wouldn't fit over the window, they called structural engineer Rich Szewczak to design a curving laminated header made of ¾-in. plywood and 2x4s.

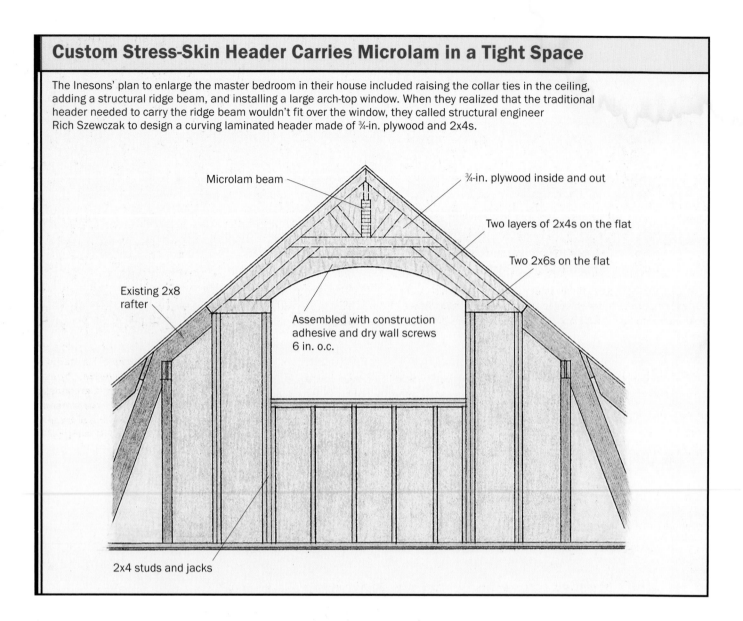

Microlam beam

¾-in. plywood inside and out

Two layers of 2x4s on the flat

Two 2x6s on the flat

Existing 2x8 rafter

Assembled with construction adhesive and dry wall screws 6 in. o.c.

2x4 studs and jacks

Careful Trim Details Unite the Exterior Elements

As we started sketching ideas for the exterior, we tried to include a number of trim elements designed to pull together the various-size window and door openings. Our intention was to lead the eye around the house, allowing it visually to align and to connect the new work to the existing house. We settled on horizontal and vertical bands of color at the window heads and jambs and below the soffits, which wrap around the shingled wall surfaces. Full 1-in.-thick cedar, stained a medium taupe color, proved to be a great material for these trim boards.

We liked the curving proportions of an existing arch-top exterior door, so we repeated that curve in several places: at the gable trim over the front door, at the trim on two rear gables, and at the window on the south gable end. This custom arch-top window (photos, facing page) aligns with an existing set of double-hung windows on the first floor. To emphasize this alignment and to provide an eye-catching break in the gable's horizontal-trim band, we added four vertical trim boards below the new window, stopping just above the existing window.

A higher ceiling opens the master bedroom. By raising the ceiling height, the authors enlarged the bedroom space, thereby making room for an arch-top casement window in the process.

Divided elements create a unified appearance. The authors created visual interest by using vertical trim to emphasize the alignment of the upper and lower windows and to break up a large area of shingles.

Our only other addition to the exterior was the attic dormer on the back of the house. Adjacent to the dormer, we installed two operable skylights that bring natural light and ventilation to the new third-floor office.

Interior Renovations Kept to a Minimum

We were satisfied with the layout of the house's interior and with the quality of the interior finishes. The master bedroom and bath, however, were cramped and in need of an update.

We raised the bedroom-ceiling height to 9 ft., which opened the room (left photo) and allowed a relatively high sill height for the new arch-top window in the south gable wall (right photo). A custom-designed casement window, the unit has the appearance of true divided lites with the advantage of low-E insulating glazing. The radius of this window arch also allowed us to position it close to the sloped ceiling.

When it became apparent that a traditional straight header would not fit above the new window, our structural engineer, Rich Szewczak, designed a stress-skin beam (drawing, facing page). Two ¾-in. plywood panels were radiused to match the curved window top and applied and secured to both sides of the 2x4 framing with construction adhesive and framing screws.

The roofline modification gave us another way to remodel the master bedroom, too. We made use of the extended space between the old and new roofs by installing recessed bookcases (left photo) as well as additional insulation.

Personal Involvement Can Mean Lower Costs and Higher Stress

Renovations are always a challenge. As owners and architects of the project, we were tested in multiples, but our involvement had its advantages, too. To help control project costs, we directly bought all major components, which were then installed by the subcontractors. We also provided our share of sweat equity; tearing out old plaster not only saved money but also raised our contractor's opinion of architects as a species.

John and Elizabeth Ineson are both architects and partners at I-Plan Architects in West Hartford, Connecticut.

Adding a Second Story

■ BY TONY SIMMONDS

It was a wonderful, prematurely warm day at the beginning of March 1994 when I first met Paul and Letizia Myers to discuss adding a second story to their house (inset photo, facing page). Both of their children were in their teens, and the house was beyond feeling cramped. A second story would give Paul and Letizia a master suite, a room for each child, and another bathroom.

That sunny March day had the kind of morning when tearing off the roof seems like the most natural and logical thing in the world. In fact, as Paul and I stood in the warm sun and looked at the roof he had repeatedly patched with elastomeric compounds, it seemed an unreasonable strain on anybody's patience to formulate a program, draw plans, and apply for permits.

In reality, the timing should have been perfect. The design could get done, and the plans drawn, in time to begin construction by late summer. August and September are the most reliably dry time of year in Vancouver.

But events foiled us. A strike at City Hall slowed the permitting process, and it was into November by the time we had approval to go ahead. Reluctantly, we shelved the project until spring. Then I met contractor Walter Ilg.

Walter makes a specialty of handling what he calls "the hard parts" of any renovation. I watched his crew remove and replace the foundation of my neighbor's house, and I was impressed with the expeditious way he handled the hard part of that one. So I showed him the plans for the Myerses' project. We agreed that the way to do it was to put up the new roof before taking down the old one. But we disagreed about timing. I had in mind the end of April. "Why wait?" he said. "It can rain anytime here."

It could do more than that, as we were to find out. But on a warm Monday in March, almost exactly a year after my first visit with the Myerses, Walter and his crew started building scaffolding.

A roof apron recalls the original proportions. Strong diagonal lines drawn by the 12-in-12 rake boards at the gable ends help to break up what would otherwise be a top-heavy facade. The lower roof continues across the front and back of the house, sheltering the windows and preserving the original roofline.

Preparing for the new roof.
The crew begins construction of the new roof by excavating post holes in the old roof over the wall plate. On the left, a ramp for removing roof debris leads to a curbside dumpster.

Prefab Trusses and Minimal Walls Help the New Roof Go Up Quickly

Walter's theory of framing is simple. You do the minimum necessary to get the roof on, throw a party, and then back-frame the rest. In this case the minimum was less than it might have been because the existing attic floor framing—2x8s on 16-in. centers—didn't have to be reinforced. Not that the job couldn't have been done the same way even if the existing joists had needed upgrading.

The new roof was also designed with minimums in mind: minimum cost and minimum delay. There would be no stick-framing; instead, factory-supplied trusses would carry the loads down the outside walls. Almost half of the trusses would be scissor trusses for the exposed wood ceiling over the stairs and in the master bedroom. The 12-in-12 pitch apron that forms the overhang at the gables and at the ground-floor eaves would be framed after the new roof was on and the exterior walls built.

To get the roof on, we needed just two bearing walls. But a continuous wall plate couldn't be installed without severing the old roof from its bearing. The solution was to use posts and beams, and to frame in the walls afterward.

Based on the layout of the interior walls, Walter and I decided to use four 4x4 posts along each side of the house. The beams would be doubled 2x10s. In one place, one beam would have to span almost 16 ft., but any deflection could easily be taken out when the permanent wall was framed underneath it. As it turned out, there wasn't any.

So on Tuesday morning, with the scaffolding built, Walter's crew cut four pockets in the appropriate locations along each side of the roof (photo, facing page). Then they secured the 4x4 posts to the existing floor framing and to the top plate of the wall below. They notched the end posts to fit into the corner made by the end joist and the rim joist. We were lucky with the intermediate ones; all of them could be fastened directly to a joist, notching the bottom of the 4x4 as required. None of the four intermediate locations was so critical, though, that the post couldn't have been moved a few inches in one direction or the other if necessary.

Walter used a builder's level to establish the height of the posts, and by Tuesday afternoon one of the beams was up and braced back to the existing roof (photo, below), and the posts were in place for the other one.

At the same time, the rest of the crew was cutting away the ridge of the existing roof to allow the flat bottom chord of the common trusses to pass across (photo, p. 144). They were able to leave the old attic collar ties/ceiling joists in place, though, because the old ceiling had been only 7 ft. 6 in. I stopped by at the end of the day to inspect the temporary post flashings the crew had made with poly and duct tape. It had been another sunny day. By afternoon, however, thin clouds had moved in, and it was getting cold. The forecast was for snow.

Posts carry a wall beam. Well braced with diagonal 2x4s, 4x4 posts rise from the holes in the roof to support a doubled 2x10 beam. Note the temporary flashings that are at the base of the posts. At the far end, the wall beam extends beyond the plane of the house to create a staging area for the roof trusses.

The crew cuts away the ridge of the existing roof to allow the flat bottom chord of the common trusses to pass across.

Snow and Rain Complicated the Job

The order for the trusses had been placed the previous week, with delivery scheduled for Thursday or Friday. But on Monday, while we were overseeing the lumber delivery, Walter let me know that he had called the truss company and promised them a case of beer if they delivered the trusses on Thursday and three cases if they got them here by Wednesday.

On Wednesday morning there was 8 in. of snow on the ground—and on the Myerses' roof. But the weather system had blown

right through, and by 8 a.m. the snow was melting fast. Walter called to say he had sent two men to sweep the snow off the roof and that the trusses would be on site if the truck could make it out of the yard. At noon I arrived to see the last bundle of trusses being landed on temporary outrigger beams.

The rest of that day was spent finishing the beam and setting and bracing the trusses. Plywood laid across where the old ridge had been scalped made it easy for one man to walk down the roof supporting the center of the truss while two others walked it along the scaffolding.

of snow but made a miserable day for the sheathing crew. Having to install the soffit, the screening, and the 2x4 purlins that tie all of the trusses together didn't speed things up. Nor did the four skylights. I didn't want the bargeboards done hurriedly, so to make things easier for the roofers, we temporarily toe-nailed 2x4s on the flat to the trimmed ends of the rake soffits. That way, the roofers could cut their shingles flush to the outside edge of the 2x4, and when the 2x4s were removed and replaced by the permanent 2x10 bargeboard and 1x3 crown, the shingles would overhang by a consistent 1¼-in. margin.

On Friday morning the roofers went to work on one side of the roof while the last nails were pounded into the sheathing on the other side. It didn't take long for them to lay the 12 squares we needed to make everything waterproof. Meanwhile, Walter and his crew were removing the old roof underneath (photo, below) and carrying it

The old roof comes down. With the new roof in place, the old one can come down. Next, the missing studs in the perimeter walls will be installed.

Even though it violated Walter's get-it-done roofing rule, I had the crew install the frieze blocking as the trusses were installed. By cutting the blocks with a chopsaw, you can ensure perfect spacing (even layout becomes unnecessary where framing proceeds on regular centers), and it's much easier to fasten the blocking this way than it is to go back and toe-nail it all afterward. Also, the soffit-venting detail I used with the exposed rafter tails required the screen to be sandwiched between two courses of soffit and stapled to the inside of the frieze block.

On Thursday another front brought wind and rain, which dispatched the last vestiges to the dumpster in 4-ft. by 12-ft. chunks. I usually try to save the old rafters, but in this case I'm afraid I let the momentum of the job dictate the recycling policy.

By 1 p.m., true to his word and to long European tradition, Walter was tying an evergreen branch to the ridge, and plates of cheese, bread, and sausage were being laid out on a sheet of plywood set up on sawhorses in the 26-ft. by 34-ft. pavilion that now occupied the top floor of the house. It might be a little breezy, as Paul said to me over a glass of wine, but at least it was dry.

A Roof Apron Prevents a Boxy Look

It took another three weeks to complete the framing and to do all of the picky work that's an inevitable part of tying everything together in a renovation. One detail, and an important element of the design, is the roof apron that encircles the house to break up the height of the building (photo, p. 141). The apron forms an eave along the front and back of the house. At the gable ends, the apron becomes a rake that rises to the peak of the roof, drawing long diagonal lines across what would otherwise be a tall, blank facade. The effect is of a 12-in-12 roof with 4½-in-12 shed dormers.

A Prefabricated Roof Apron

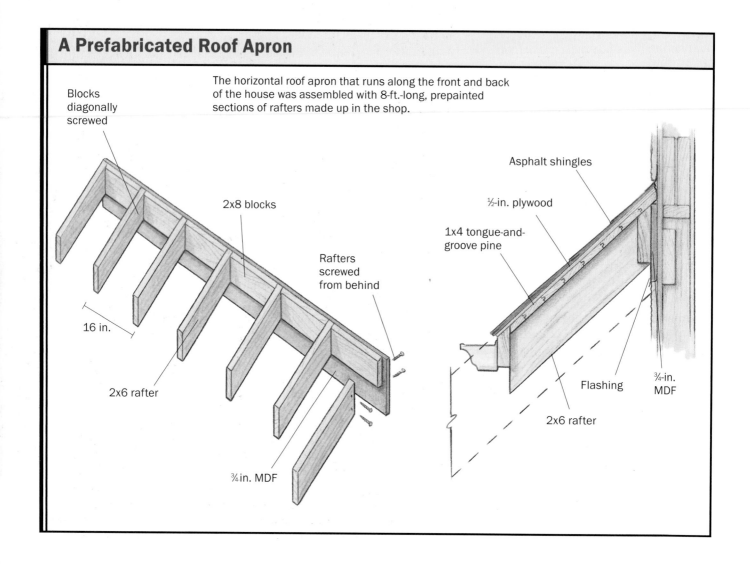

Blocks diagonally screwed

The horizontal roof apron that runs along the front and back of the house was assembled with 8-ft.-long, prepainted sections of rafters made up in the shop.

2x8 blocks

Rafters screwed from behind

16 in.

2x6 rafter

¾-in. MDF

Asphalt shingles

½-in. plywood

1x4 tongue-and-groove pine

Flashing

¾-in. MDF

2x6 rafter

The apron has practical value, too, particularly at the eave, where it covers the top edge of the existing wall finish, providing an overhang to protect the ground-floor windows. If you're building outside the painting season, it's essential to get a coat of paint on everything before it's applied to the outside of the house, so we built as much as we could of this apron in 8-ft. sections in my shop (drawing, facing page). For example, the eaves consist of 2-ft. long 2x6 lookout rafters screwed from the back to a 12-in.-wide strip of Medex®, an exterior-grade medium-density fiberboard that is gaining popularity for use as exterior trim here. Frieze blocks cut from 2x8s act as pressure blocks between the rafters. We prepainted these assemblies and the 1x4 tongue-and-groove pine that we nailed to their tops in our shop.

On site, the eave sections were installed and tied together with the prepainted 1x4s and 2x6 fascia. Then we snapped lines on the gable ends from the ridge to the eave lookouts to establish the line of the rake soffit (photo, right). On this line, we toenailed a triangular bump-out, framed out of 2x10s, to the gable-wall framing. From the base of the bump-out, we ran a 2x6 that acts as a rake trim board for most of its length and then becomes the last lookout rafter where it runs into the eave overhang.

We nailed preassembled and prepainted strips of soffit to the rake trim and to the gable bump-out. Made of tongue-and-groove 1x4s blind-nailed to 18-in.-wide strips of ½-in. plywood, the 8-ft.-long strips of rake soffit were pretty floppy until the 2x10 bargeboards went on.

Projecting the gable peaks out from the plane of the wall did more than provide solid support for the rake apron with its heavy bargeboard. It also created some visual interest and gave a little protection to the bedroom windows in the east wall. The peaks were finished with louvered vents and 1x4 bevel siding. These peaks make a nice big triangle of painted woodwork to balance the large areas of stucco.

Bump-out and fascias support the rake soffit. At the gable end, a bump-out protects the upstairs windows and supports the tops of the 2x10 bargeboards. The 2x10s are borne by 2x6 fascias cantilevered past the roof-apron rafters. Note how a built-up water table makes a clean line between the old stucco and the new.

We also ran a water table at the second-floor joist level (photo, p. 147). Besides its aesthetic contribution, this band covers the flashing protecting the top edge of the old stucco and makes a practical separation so that new stucco and old don't have to meet. Detailing woodwork so that stucco always has a place to stop and so that no one panel of it is too big makes the plasterer's job a whole lot easier.

Shaun Friedrich, who learned the stucco trade from his father and who can tell without leaving his truck what a particular stucco is, when it was done, and quite often who did it, made a beautiful job of approximating the look of the original dry-dash finish. Dry dash is a labor-intensive stucco finish in which a layer of small, sharp stones is embedded in a layer of mortar. Shaun rendered a compatible finish for the upstairs walls by using a drywall-texturing gun to create the random, splattered look of dry dash. This substitution saved us $1,000.

Allocating the New Space

On the inside, Walter's crew was turning the 26-ft. by 34-ft. pavilion into a second floor with three bedrooms and two baths (floor plan, below). The west end contains a master bedroom and bath. In the center of the house, a hallway includes the existing stair, a bathroom at the north end (photo, facing page) and a balcony at the south end. Bedrooms at the east end complete the plan.

Injecting Variety into a Rectangular Plan

Nooks, alcoves, skylights, and dropped ceilings all play their part in enlivening the plan.

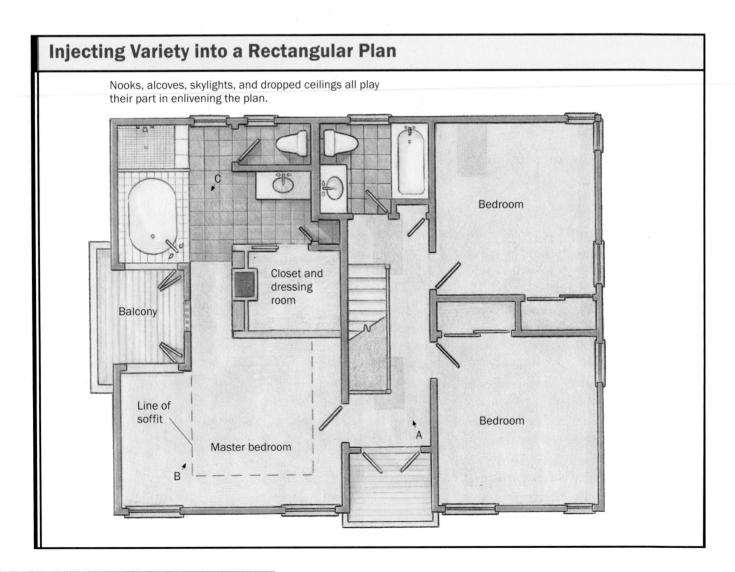

Balcony

Line of soffit

Master bedroom

Closet and dressing room

Bedroom

Bedroom

Daylight in the center of the house. Skylights over the centrally located hallway light up the stairs, as well as the bathroom, by way of its generous transom. Photo taken at A on floor plan.

The subdivision of the master-bedroom space to accommodate a walk-in closet and the bathroom was the most intriguing part of the design. I wanted the room to feel large and generously proportioned, but at the same time I wanted the different areas within it to be well defined.

The first division is between north and south. The bathroom, with its requirement for privacy, is on the north side; the bedroom is on the south. What separates them is not a wall but other subsidiary spaces: the walk-in closet and a small balcony (photo, p. 150).

Then there is the division between the main part of the room and the three 6-ft.-deep alcoves along the west wall. Linked by their common ceiling height—7 ft.—the alcoves contain from north to south the shower/tub space; the balcony; and the bedroom-dresser area.

A balcony separates the bedroom and the bath. Along the west wall, three alcoves with low ceilings have distinct functions. In the foreground, the shower and tub occupy the first alcove. In the middle, a small balcony overlooks the secluded backyard. In the distance, the third alcove provides space for the bedroom dresser. Photo taken at C on floor plan.

In addition to their common ceiling heights, the alcoves are further linked by large windows that open onto the balcony (photo, right). These windows can be folded back against the wall so that in nice weather the balcony is really a part of the bedroom.

The transparency of these linked alcoves to one another goes a step further. On the bathroom side, the shower is separated from the tub by a glass partition; on the bedroom side a window in the south wall lines up with the two windows to the balcony. Standing in the shower, you can look right through four transparent layers to the outside. In a small house, long views such as these foster a sense of spaciousness.

The ceiling in the master bedroom is an example of how to turn a technical problem to practical advantage. The decision to use trusses throughout for the sake of expeditiousness and economy meant that the ceiling could slope only at a pitch of 2-in-12 (the bottom chord of a 4½-in-12 scissor truss), and that skylight wells would necessarily be rather deep. Locating the skylight so that the slope of one side of the ceiling extends into the skylight well to meet the top of the skylight makes for a dramatic light shaft that spills light all along the ceiling as well as down the wall (photo, right). It also didn't leave much room for error in the layout we had to do back on that raw day in March when Walter's crew members were swarming over the roof with snow in their hair and shinglers at their heels.

As for the low slope of the ceiling, we made it seem higher by holding the closet walls to a height of 7 ft. In the end the effect was everything we had hoped it would be. Letizia, who is Swiss and for whom I was trying to echo a wooden chalet ceiling, was not disappointed.

Tony Simmonds has been involved with woodworking and building for over 25 years. He operates DOMUS, a design/build firm in Vancouver, British Columbia.

The outdoors is nearby. On the left, folding windows lead to a balcony off the master suite. On the right, a 7-ft. closet wall separates bedroom from lavatory. The sloping ceiling extends beyond the ridge to become part of the skylight well over the closet. Photo taken at B on floor plan.

CREDITS

The articles in this book appeared in the following issues of *Fine Homebuilding*.

p. iii: Photo by Steve Culpepper, courtesy of *Fine Homebuilding*, © The Taunton Press, Inc.

p. iv: (left) Photo by Steve Culpepper, courtesy of *Fine Homebuilding*, © The Taunton Press, Inc.; (right) Photo by Charles Miller, courtesy of *Fine Homebuilding*, © The Taunton Press, Inc.

p. v: (left) Photo © Nicholas Pitz; (right) Photo © Rich Ziegner.

p. 4: An Addition Foundation by Rick Arnold, issue 145. Photos by Roe A. Osborn, courtesy of *Fine Homebuilding*, © The Taunton Press, Inc., except photos on p. 9 (bottom) and p. 11 © Rick Arnold.

p. 13: Supporting an Addition by Philip S. Wenz, issue 98. Photos © Philip S. Wenz; drawings by Vince Babak, courtesy of *Fine Homebuilding*, © The Taunton Press, Inc.

p. 20: Laying Up Concrete Block by John Carroll, issue 111. Photos by Steve Culpepper, courtesy of *Fine Homebuilding*, © The Taunton Press, Inc.

p. 32: A Builder's Screen Porch by Scott McBride, issue 86. Photos © Scott McBride except photos on p. 32 (bottom) and p. 40 by Jefferson Kolle, courtesy of *Fine Homebuilding*, © The Taunton Press, Inc.; drawings by Bob Goodfellow, courtesy of *Fine Homebuilding*, © The Taunton Press, Inc.

p. 43: A Dining Deck by Tony Simmonds, issue 93. Photos by Charles Miller, courtesy of *Fine Homebuilding*, © The Taunton Press, Inc. except photo on p. 46 © Tony Simmonds; drawings by Bob La Pointe, © The Taunton Press, Inc.

p. 50: Adding a Sunroom with Porch by Didier Ayel, issue 114. Photos by Steve Culpepper, courtesy of *Fine Homebuilding*, © The Taunton Press, Inc., except for inset photo on p. 51 © Didier Ayel; drawings by Dan Thorton, courtesy of *Fine Homebuilding*, © The Taunton Press, Inc.

p. 56: A Classic Sunroom by Steven Gerber, issue 107. Photos by Roe A. Osborn, courtesy of *Fine Homebuilding*, © The Taunton Press, Inc.; Drawings by Dan Thorton, courtesy of *Fine Homebuilding*, © The Taunton Press, Inc.

p. 62: Keeping a Dormer Addition Clean and Dry by Nicholas Pitz, issue 158. Photos © Nicholas Pitz; drawings © Don Mannes.

p. 68: A Different Approach to Rafter Layout by John Carroll, issue 115. Photos by Steve Culpepper, courtesy of *Fine Homebuilding*, © The Taunton Press, Inc., except photo on p. 73 by Scott Phillips, courtesy of *Fine Homebuilding*, © The Taunton Press, Inc.; drawings by Dan Thornton, courtesy of *Fine Homebuilding*, © The Taunton Press, Inc.

p. 78: A Gable-Dormer Retrofit by Scott McBride, issue 134. Photos © Scott McBride; drawings by Christopher Clapp, courtesy of *Fine Homebuilding*, © The Taunton Press, Inc.

p. 88: Supporting a Cantilevered Bay by Mike Guertin, issue 136. Photos by Roe A. Osborn, courtesy of *Fine Homebuilding*, © The Taunton Press, Inc.

p. 94: Remodeling with Metal Studs by Tom O'Brien, issue 97. Photos by Roe A. Osborn, courtesy of *Fine Homebuilding*, © The Taunton Press, Inc.

p. 104: Solo Drywall Hanging by Pat Carrasco, issue 140. Photos by Tom O'Brien, courtesy of *Fine Homebuilding*, © The Taunton Press, Inc.

p. 114: A Dramatic Family-Room Addition by B. Alan Leonard, issue 97. Photos © B. Alan Leonard except photo on p. 115 © Rich Ziegner; drawings by Vince Babak, courtesy of *Fine Homebuilding*, © The Taunton Press, Inc.

p. 122: A Well-Lit Addition by Robert L. Marx, issue 83. Photos © Robert L. Marx; drawings by Christopher Clapp, courtesy of *Fine Homebuilding*, © The Taunton Press, Inc.

p. 127: Converting a Garage into Living Space by Neil Hartzler, issue 77. Photos on p. 128 (top) by Neil Hartzler; Photos on p. 128 (bottom) and p. 133 © Michael Fulks; drawing by Vince Babak, courtesy of *Fine Homebuilding,* © The Taunton Press, Inc.

p. 134: A Suburban Metamorphosis by John and Elizabeth Ineson, issue 113. Photos by Charles Bickford, courtesy of *Fine Homebuilding,* © The Taunton Press, Inc., except p. 135 (bottom) © John Ineson; drawings by Vince Babak, courtesy of *Fine Homebuilding,* © The Taunton Press, Inc.

p. 140: Adding a Second Story by Tony Simmonds, issue 102. Photos © Tony Simmonds except p. 141 (top), p. 147, p. 149 by Charles Miller, courtesy of *Fine Homebuilding,* © The Taunton Press, Inc.; drawings by Bob La Pointe, courtesy of *Fine Homebuilding,* © The Taunton Press, Inc.

INDEX